EXPEDITION ZUKUNFT
SCIENCE EXPRESS

WIE WISSENSCHAFT UND TECHNIK UNSER LEBEN VERÄNDERN
HOW SCIENCE AND TECHNOLOGY CHANGE OUR LIFE

Expedition Zukunft – Science Express
Wie Wissenschaft und Technik unser Leben verändern
How Science and Technology Change our Life

IMPRESSUM *imprint*

ISBN 978-3-534-23396-0
November 2009
Expedition Zukunft – Science Express online: www.expedition-zukunft.org

Dieser Katalog begleitet den Wissenschaftszug *Expedition Zukunft*, der durch ein Projektteam der Max-Planck-Gesellschaft zur Förderung der Wissenschaften e. V., München, Deutschland, konzipiert und umgesetzt wurde. *This catalogue accompanies the* Science Express, *a travelling multimedia exhibition conceived and realised by a project team of the Max Planck Society for the Advancement of Science, Munich, Germany.*

HERAUSGEBER *publisher* Max-Planck-Gesellschaft zur Förderung der Wissenschaften e. V., Stabsreferat für Forschungsanalyse und -vorausschau. © 2009. Die Urheberrechte der Abbildungen liegen bei den Wissenschaftlern bzw. bei ihren Instituten. *Max Planck Society for the Advancement of Science, Office of Research Analysis and Foresight. © 2009. The copyright of the figures remains with the scientists or their institutes.*

PROJEKTTEAM EXPEDITION ZUKUNFT *Project Team Science Express*
Hofgartenstr. 8, 80539 München
Telefon +49 (89) 2108 - 1407   Fax +49 (89) 2108 - 1243
expedition-zukunft@gv.mpg.de   www.mpg.de

DANKSAGUNG *acknowledgements* Die Max-Planck-Gesellschaft dankt allen am Katalog beteiligten Wissenschaftlern und Instituten sowie allen Wissenschafts-, Wirtschafts- und Medienpartnern des Zuges für die großzügige Überlassung von Bildrechten. Wir bitten um Verständnis, dass aus Platzgründen in der Ausstellung nicht alle Beteiligten aufgeführt werden können. *The Max Planck Society expresses its thanks to all scientists and institutes contributing to the Science Express exhibition catalogue as well as to its scientific, corporate and media partners for their generous image copyright permissions. Due to the large number of contributors, we are not able to mention everyone. We do however wish to thank each and every one of them.*

REDAKTION *editorial office* Nadja Pernat, Projektteam Expedition Zukunft
KLÄRUNG BILDRECHTE *copyright clearance* Melanie Kuhnle
ÜBERSETZUNGEN *translations* Baker & Harrison, München
FOTOGRAFIE *photography* Oliver Wia, Berlin
WEITERE FOTOS VON *additional photography by* Frank Peters, Daniel Sobotta, Hans Herbig, Dawin Meckel, Ulrike Richter, Meike Jotzo, Dr. Clemens Schneeweiß
DRUCK *printing* Medialis Offsetdruck GmbH, Berlin
BINDUNG *bookbinding* Leipziger Verlags- und Industriebuchbinderei GmbH, Leipzig

Die Max-Planck-Gesellschaft dankt allen Partnern und Förderern, allen beteiligten Wissenschaftlern und Leihgebern sowie den 62 gastgebenden Städten für die großzügige Unterstützung des Ausstellungszuges *Expedition Zukunft*.

The Max Planck Society would like to thank all partners and supporters, all scientists and institutions involved, as well as the 62 host cities for their generous support of the *Science Express* exhibition train.

WISSENSCHAFTSPARTNER *scientific partners*

Deutsche Forschungsgemeinschaft
DFG

DEUTSCHE UNIVERSITÄTEN & HOCHSCHULEN

WIRTSCHAFTSPARTNER *corporate partners*

ZUGPARTNER *train partners*

WAGENPARTNER *carriage partners*

MITMACHLABOR *hands-on laboratory*

Stifterverband
für die Deutsche Wissenschaft

MEDIENPARTNER *media partners*

Spektrum
DER WISSENSCHAFT   [W] wie Wissen®

Vorworte *Prefaces*	7
Einleitung *Introduction*	13
MAKING-OF *making-of* Eine Ausstellung entsteht *An exhibition comes into being*	19
EXPEDITION ZUKUNFT *science express* Was kommt auf uns zu? *What are we facing?*	45
WOHER + WOHIN *where from + where to* Die Suche nach den Ursprüngen *The search for our origins*	53
BIO + NANO *bio + nano* Nano- und Biowissenschaften verschmelzen *The convergence of nanoscience and bioscience*	69
INFO + KOGNO *info + cogno* Das Gehirn – ein intelligenter Computer? *The brain – an intelligent computer?*	87
VERNETZT + GLOBAL *networked + global* Auf dem Weg in eine digitale Gesellschaft *On the road to a digital society*	103
INTELLIGENT + VIRTUELL *intelligent + virtual* Neue Materialien und die Produktion der Zukunft *Innovative materials and the factory of the future*	123
HINTER DEN KULISSEN *behind the scenes* Eine kleine Welt für sich *A small world of its own*	139
WIRKSAM + INDIVIDUELL *effective + individual* Wird es eine Welt ohne Krankheiten geben? *Will we ever have a disease-free world?*	157
GESUND + PRODUKTIV *healthy + productive* Wie werden wir neun Milliarden Menschen ernähren? *How will we feed nine billion people?*	175
NACHHALTIG + EFFIZIENT *sustainable + efficient* In Kreisläufen denken – Ressourcen schonen *Thinking in cycles – saving resources*	193
FLEXIBEL + DIGITAL *flexible + digital* Unterwegs Zuhause: Mobilität und modernes Leben *At home on the move: Mobility and modern life*	211
NATÜRLICH. KÜNSTLICH *natural. artificial* Die Zukunft des Menschen *The future of mankind*	231
ENTDECKEN + STAUNEN *discover + marvel* Die Zukunft gestalten *Shaping the future*	251
SIE WAREN AM ZUG *aboard the train* Zahlen und Fakten einer außergewöhnlichen Tournee *Facts and figures of a remarkable tour*	261

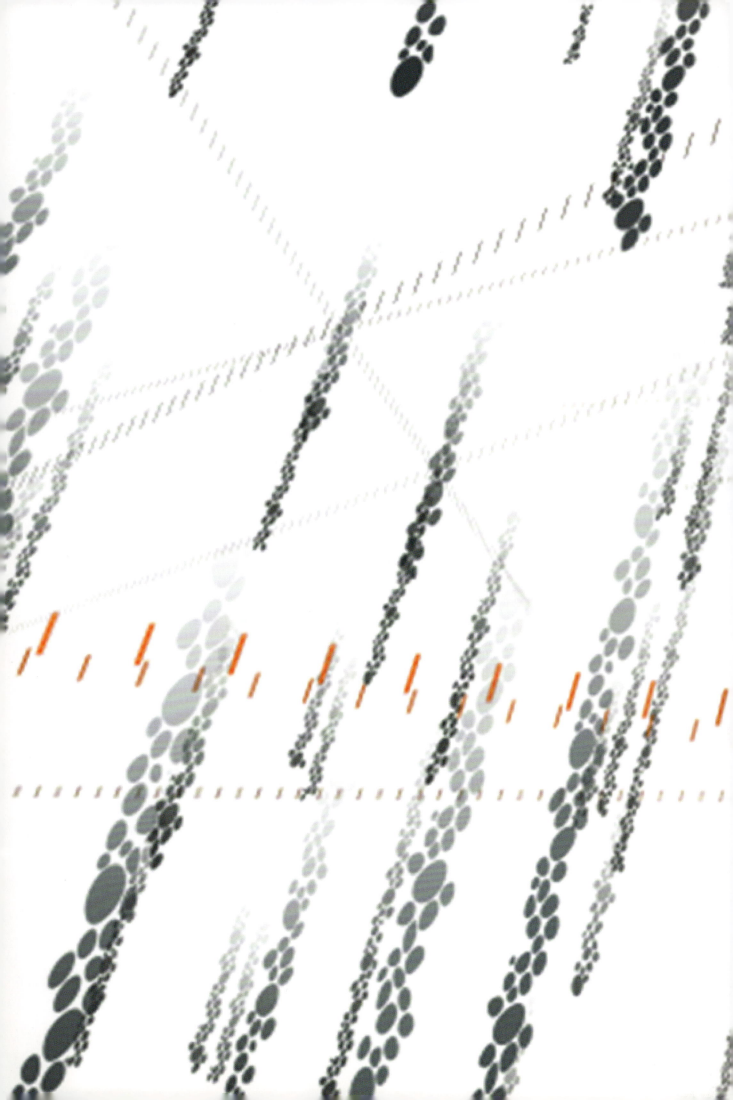

Im *Wissenschaftsjahr 2009* habe ich Sie zu einer Entdeckungsreise in die deutsche Wissenschaftslandschaft eingeladen: der *Forschungsexpedition Deutschland*. Eine Vielzahl von Partnern aus Wissenschaft, Wirtschaft und Kultur machte in Veranstaltungen, Wettbewerben und Ausstellungen Wissenschaft erlebbar. Der Ausstellungszug *Expedition Zukunft* präsentierte in zwölf Waggons die vielen Facetten von Wissenschaft und Forschung in Deutschland und zeigte anschaulich die Perspektiven auf, die uns in den nächsten zehn Jahren durch sie ermöglicht werden. Insbesondere Schülerinnen und Schüler hat der Ausstellungszug für Forschung fasziniert. Denn die Wissenschaft bietet nachfolgenden Generationen einen spannenden Auftrag – nämlich der Welt von übermorgen den Weg zu bereiten. ——Wir müssen in der Gesellschaft ein Klima schaffen, in dem die Arbeit von Forscherinnen und Forschern wieder mehr Wertschätzung erfährt. Denn gerade sie schaffen die Quellen unseres künftigen Wohlstands. Deshalb müssen wir Forschung und Wissenschaft stärken, um im weltweiten Innovationswettbewerb bestehen zu können.

Prof. Dr. Annette Schavan, MdB
Bundesministerin für Bildung und Forschung

In the *Year of Science 2009*, I invited everybody to embark on a voyage of discovery through the German scientific landscape – to join the *Research Expedition Germany*. A wide range of partners from the worlds of science, business and culture have made science come alive in their many events, contests and exhibitions. The *Science Express* exhibition train has been presenting the many facets of science and research in Germany in its twelve rail cars, clearly illustrating the perspectives that will be opened up to us in the next ten years. The exhibition train aimed to inspire schoolchildren in particular to become excited about research. After all, science offers coming generations a fascinating task – that of paving the way for the world of tomorrow. ——We must create a climate in our society in which the work of scientists is held in greater esteem. For it is scientists who create the sources of our future wealth. That is why it is essential for us to strengthen research and science in order to survive in the global competition for innovation.

Prof. Annette Schavan, MdB
Federal Minister of Education and Research

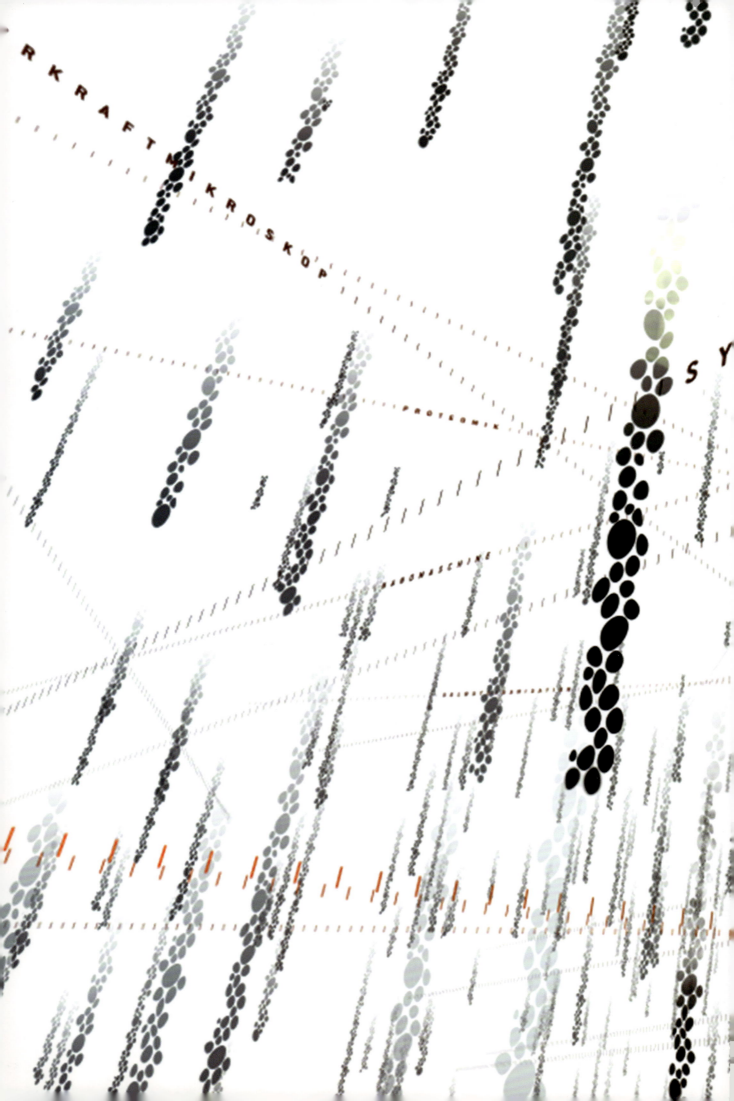

Die Welt um uns herum befindet sich in einem rapiden Umbruch. Forschung und Entwicklung werden inzwischen rund um den Globus als Motor für Wachstum und Entwicklung und als Katalysator sozialer Veränderungen im 21. Jahrhundert anerkannt. Was heute weltweit erforscht wird, prägt künftig unser Leben. Wir wissen: Dem Anwenden geht das Erkennen voraus. Wir müssen uns heute an möglichst vielen Forschungsthemen beteiligen, damit unseren Kindern genügend Erkenntnisse zur Verfügung stehen, aus denen sie Innovationen für den Wohlstand und das Wohlergehen heutiger wie kommender Generation sichern können. ___ Die *Expedition Zukunft* vermittelt einen Eindruck, welche neuen Möglichkeiten durch Wissenschaft und Forschung in den kommenden Jahren eröffnet werden. Die Zukunft ist nah, sie wird viel Neues bringen – wir müssen sie heute gestalten.

Prof. Dr. Peter Gruss
Präsident der Max-Planck-Gesellschaft zur Förderung der Wissenschaften e. V.

The world around us is undergoing a rapid transformation. Research and development are now recognized across the globe as an engine of growth and development and a catalyst of societal change in the 21st century. The topics being researched today will shape our lives into the future. For as we are keenly aware, knowledge precedes application. We need to be involved in as many research topics as we possibly can today so that our children will have sufficient knowledge at their disposal to enable them to secure innovations for the wealth and wellbeing of their own and future generations. ___ The *Science Express* conveys a feeling of the new possibilities that science and research will open up in the years to come. The future is near, it will bring much that is new – and we must shape the future today.

Prof. Peter Gruss
President of the Max Planck Society for the Advancement of Science

EXPEDITION
SCIENCE

Fragen Sie sich manchmal auch, wie die Welt in zwanzig Jahren aussehen wird? Wie wir dann wohl leben werden? Mit dem Ausstellungszug *Expedition Zukunft* möchte die Max-Planck-Gesellschaft allen Besuchern und speziell jungen Menschen einen Überblick davon vermitteln, welche Wissensgebiete sich heute weltweit besonders dynamisch und vielversprechend entwickeln und wie diese in den kommenden zwei Jahrzehnten unser Leben beeinflussen. Doch der Wissenschaftszug macht Forschung und Entwicklung nicht nur jungen Menschen als beruflichen Weg schmackhaft. Die *Expedition Zukunft* erzählt davon, dass wir die modernen Natur- und Lebenswissenschaften brauchen, um uns persönlich wie unserer ganzen Zivilisation eine gute Zukunft zu sichern. —— Mit einem Anteil von bald drei Prozent der nationalen Wertschöpfung sind Wissenschaft und Forschung nicht nur in Deutschland ein eigener Wirtschaftszweig. Geforscht wird längst im internationalen Wettbewerb. Die Themen sind nicht nur für sich allein genommen sehr spannend, ihre Ergebnisse werden auch dazu beitragen, einem nachhaltigen Wirtschafts- und Gesellschaftsmodell zum Durchbruch zu verhelfen. Wissenschaft und Technologie spielen in einem rohstoffarmen Land wie Deutschland eine ganz besondere Rolle. Erst recht, wenn in der Zukunft viele Naturressourcen der Erde zur Neige gehen und nicht mehr den Bedarf einer wachsenden Weltbevölkerung befriedigen können. Das betrifft nicht nur Erdöl, Erdgas, Nahrungsmittel oder Trinkwasser, sondern auch Erze, Salze für Düngemittel oder medizinisch relevante Naturstoffe. —— Die *Expedition Zukunft* entführt ihre Besucher in die Welt von morgen, informiert vorausschauend über Themen und Entwicklungen, die gerade erst im Entstehen sind. Es geht also nicht darum, bereits vorhandenes Wissen besser zu erläutern oder bestimmte Technologien zu rechtfertigen. Es geht um Entwicklungstrends und -möglichkeiten, die sich erst andeuten und denen wir uns stellen müssen. Die Zukunft wird heute gestaltet. —— Wie ein begehbares Buch werden die Themen des Zuges erzählt. Nach dem Prolog im ersten Wagen wird man in den nächsten drei Wagen auf Entwicklungen in der Grundlagenforschung eingestimmt, die allen anderen Bereichen zusätzliche Impulse verleihen. In den folgenden sechs Wagen erfährt man mehr darüber, wie sich Medizin und Ernährung, unsere Lebensweise, unsere Arbeit und auch unsere Art zu kommunizieren, verändern werden. Digitalisierung, Miniaturisierung und Personalisierung sind Stichworte, die den komplexen Wandel bei Produkten und Leistungen beschreiben. Der nächste Wagen lädt dann zum Nachdenken darüber ein, welche neuen Chancen Wissenschaft und Forschung eröffnen, aber auch welche Rahmenbedingungen erforderlich sein müssen, damit dieses Potenzial möglichst allen Menschen zugute kommt. Schließlich gibt es im letzten Wagen ein Mitmachlabor, in dem die Besucher beim Experimentieren staunen und ihren Forschergeist entdecken können. —— Der Ausstellungszug als Teil des *Wissenschaftsjahres 2009 – Forschungsexpedition Deutschland* erreichte viele Menschen flächendeckend und in kurzer Zeit. Das *Wissenschaftsjahr 2009* wird vom Bundesministerium für Bildung und Forschung und der Initiative *Wissenschaft im Dialog* gemeinsam mit der Deutschen Akademie der Naturforscher Leopoldina, der Robert Bosch Stiftung und dem Stifterverband für die Deutsche Wissenschaft ausgerichtet. —— Nicht nur durch die Präsentation in einem Zug, sondern auch in der Art der Gestaltung beschritt die Ausstellung neue Wege. Ziel war es, eine Ausstellung zu schaffen, die ein breites Publikum für sehr anspruchsvolle wissenschaftliche Inhalte zu begeistern vermag. Die Gestaltung hatte die Vielfalt zwölf unterschiedlicher Themen ebenso zu berücksichtigen wie die extreme Raumsituation eines über 300 Meter langen und nur 2,80 Meter breiten Zuges. Jeder der zwölf Wagen erhielt ein eigenes Design, welches das jeweilige Thema in assoziativer und emotionaler Weise aufnimmt

und Zukunftsszenarien schafft, die durch ihre Atmosphäre und Lebendigkeit den Besucher in Bann ziehen. Die komplexen Inhalte der Ausstellung werden in einer Kombination aus Multimedia, interaktiven Exponaten, spektakulären Originalobjekten aus der Forschung und ein komplettes Mitmachlabor vermittelt. ▬ Die Ausstellung zeigt so einen Schnappschuss – eine Momentaufnahme aktueller Forschung in Laboren und Experimentierhallen – und macht den Besucher zum Zeugen einer Entwicklung, deren Dynamik sonst meist verborgen bleibt. Sie zeigt den Puls der Zeit, dem sich kein Land, keine Region, keine Stadt längerfristig entziehen kann. Diese Inszenierung musste in der extrem kurzen Zeit von neun Monaten realisiert werden. Die Herausforderung bestand aber auch darin, für die schwierigen technischen Bedingungen, die vor allem durch die mobilen Räume, ein technisch begrenztes Stromkontingent und die Notwendigkeit der Raumklimatisierung gekennzeichnet waren, ausstellungsgerechte Lösungen zu finden. Idee und Konzept der Ausstellung wurden von einem Projektteam der Max-Planck-Gesellschaft entwickelt; Design und Szenographie stammen von der Agentur ArchiMeDes, die die Ausstellung auch gebaut hat. ▬ Seit Oktober 2007 ist ein Zug mit der ebenfalls von der Max-Planck-Gesellschaft konzipierten Ausstellung *Science Tunnel* in Indien unterwegs – der überaus große Erfolg des Projekts legte es nahe, auch in Deutschland eine Wissenschaftsausstellung auf Schienen durch das Land zu schicken. Entstanden ist hierfür eine komplett neue Ausstellung, die einen Überblick über die Forschung und Entwicklung in ganz Deutschland gibt. ▬ Die Ausstellung *Expedition Zukunft* wurde hierfür von vielen Partnern aus Wissenschaft und Wirtschaft mit Ausstellungsstücken und Inhalten unterstützt. Neben der Max-Planck-Gesellschaft beteiligten sich die Fraunhofer-Gesellschaft, die Helmholtz-Gemeinschaft, die Leibniz-Gemeinschaft, sowie zahlreiche Universitäten und Hochschulen, die Unternehmen Bayer AG, Siemens AG und Volkswagen AG als Zugpartner, und als Wagenpartner BASF SE, Robert Bosch GmbH, Deutsche Bahn AG, Deutsche Telekom AG, OSRAM GmbH, der Stifterverband für die deutsche Wissenschaft und der Verband Forschender Arzneimittelhersteller e.V. ▬ Wir danken dem Bundesministerium für Bildung und Forschung und allen Partnern. Erst ihre Unterstützung hat es uns ermöglicht, die *Expedition Zukunft* auf den Weg zu bringen.

Projektteam Expedition Zukunft der Max-Planck-Gesellschaft
Jan Bejenke, Dalija Budimlic, Christoph Ettl, Hannelore Hämmerle, Nadja Pernat, Peter M. Steiner, Andreas Trepte, Christiane Walch-Solimena

Do you ever ask yourself what the world will look like twenty years from now? And how our lives will be then? With the *Science Express* exhibition train the Max Planck Society wants to give all visitors, and young people in particular, an insight into the scientific fields that are evolving on a global scale in a particularly dynamic and promising way, and how they will influence our lives in the coming two decades. However, the aim of the science train is not just to make research and development attractive to young people as a future career option. The *Science Express* shows how we need today's natural and life sciences to ensure a positive future, both for ourselves and for our entire civilization. Accounting for almost three percent of Germany's gross domestic product, science and research constitute an independent sector of the economy – not just in Germany. Research has long been carried out in the context of international competition. Not only are the research topics very exciting in themselves, the results achieved will also help in making the breakthrough to a sustainable economic and social model. Science and technology have an important role to play in a country poor in natural resources like Germany. This will be even more true in the future when many of the Earth's natural resources run low and are no longer able to meet the needs of a growing global population. The resources concerned are not only the obvious ones like oil, natural gas, food and drinking water, but others such as mineral ores, salts for fertilizers and medically relevant natural materials as well. —— The *Science Express* transports its visitors into the world of tomorrow and provides forward-looking information about topics and developments that are only just emerging. It is not concerned with better explaining information that is already available or justifying certain technologies. The focus is on emerging trends and possibilities that are now coming into view and that must be dealt with. The future is being shaped today. —— The topics covered by the exhibition are narrated in the style of a walk-through book. After a prologue in the first carriage, the next three coaches tell visitors about recent developments in basic research, which provide inspiration for the work carried out in other areas. In the next six carriages, visitors learn more about how medicine and nutrition, our lifestyle, our work and our ways of communication will change in the years to come. Digitization, miniaturization and personalization are some of the keywords describing the complex changes taking place in products and services. The following carriage then invites visitors to consider the new opportunities offered by science and research, as well as the conditions necessary to ensure that, wherever possible, this potential can be of benefit to all. Finally, in the last carriage there is a hands-on laboratory where visitors can marvel at their own discoveries. —— As part of the *Year of Science 2009 – Research Expedition Germany*, the exhibition train reached a large number of people across the entire country. The *Year of Science 2009* is hosted by the Federal Ministry of Education and Research and the *Science in Dialog* Initiative, in cooperation with the German Academy of Sciences Leopoldina, the Robert Bosch Foundation and the Donors' Association for the Promotion of Sciences and Humanities in Germany. —— However, it was not only the presentation in a train that was new, but also the design stroke a new path. The goal was to create an exhibition that could inspire a wide audience for very sophisticated scientific content. The design had to allow for twelve very different topics as well as the extreme spatial conditions of a 300 metres long and only 2.80 metres wide train. Each of the twelve carriages obtained its own design, which took up the respective topics in an associative and emotional way and created future scenarios that capture the visitors' attention by their atmosphere and vibrancy. The complex content of the exhibition is communicated through a combination of multimedia displays, interactive exhibits,

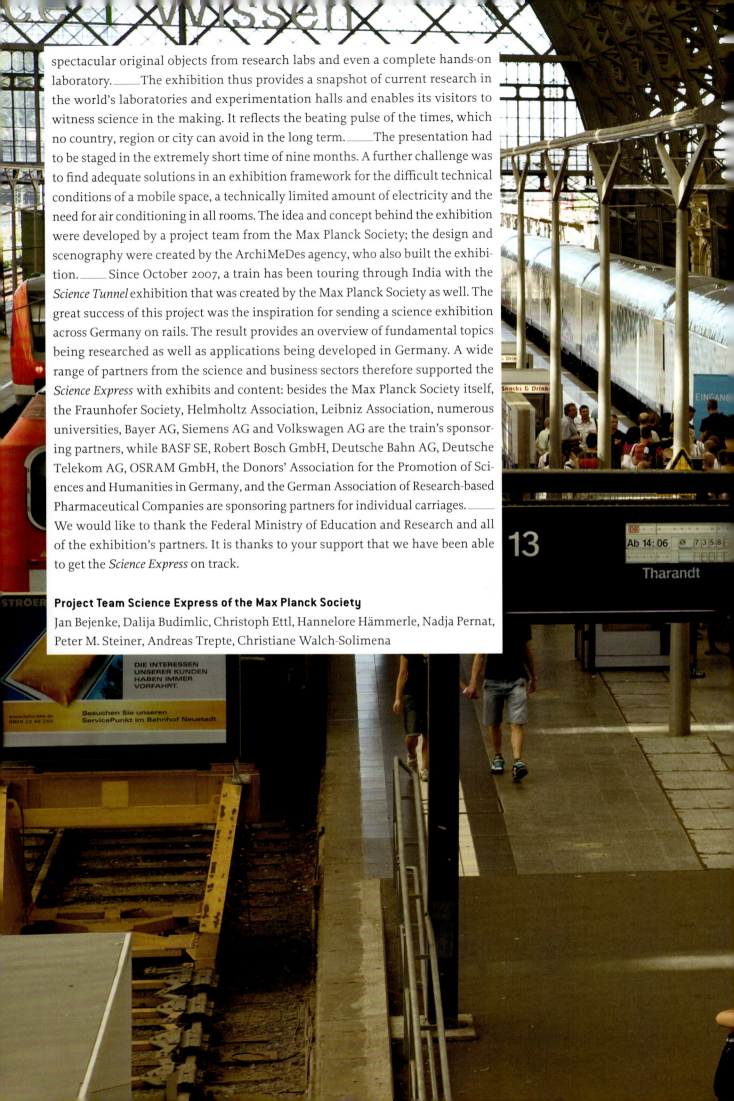

spectacular original objects from research labs and even a complete hands-on laboratory. —— The exhibition thus provides a snapshot of current research in the world's laboratories and experimentation halls and enables its visitors to witness science in the making. It reflects the beating pulse of the times, which no country, region or city can avoid in the long term. —— The presentation had to be staged in the extremely short time of nine months. A further challenge was to find adequate solutions in an exhibition framework for the difficult technical conditions of a mobile space, a technically limited amount of electricity and the need for air conditioning in all rooms. The idea and concept behind the exhibition were developed by a project team from the Max Planck Society; the design and scenography were created by the ArchiMeDes agency, who also built the exhibition. —— Since October 2007, a train has been touring through India with the *Science Tunnel* exhibition that was created by the Max Planck Society as well. The great success of this project was the inspiration for sending a science exhibition across Germany on rails. The result provides an overview of fundamental topics being researched as well as applications being developed in Germany. A wide range of partners from the science and business sectors therefore supported the *Science Express* with exhibits and content: besides the Max Planck Society itself, the Fraunhofer Society, Helmholtz Association, Leibniz Association, numerous universities, Bayer AG, Siemens AG and Volkswagen AG are the train's sponsoring partners, while BASF SE, Robert Bosch GmbH, Deutsche Bahn AG, Deutsche Telekom AG, OSRAM GmbH, the Donors' Association for the Promotion of Sciences and Humanities in Germany, and the German Association of Research-based Pharmaceutical Companies are sponsoring partners for individual carriages. —— We would like to thank the Federal Ministry of Education and Research and all of the exhibition's partners. It is thanks to your support that we have been able to get the *Science Express* on track.

Project Team Science Express of the Max Planck Society
Jan Bejenke, Dalija Budimlic, Christoph Ettl, Hannelore Hämmerle, Nadja Pernat, Peter M. Steiner, Andreas Trepte, Christiane Walch-Solimena

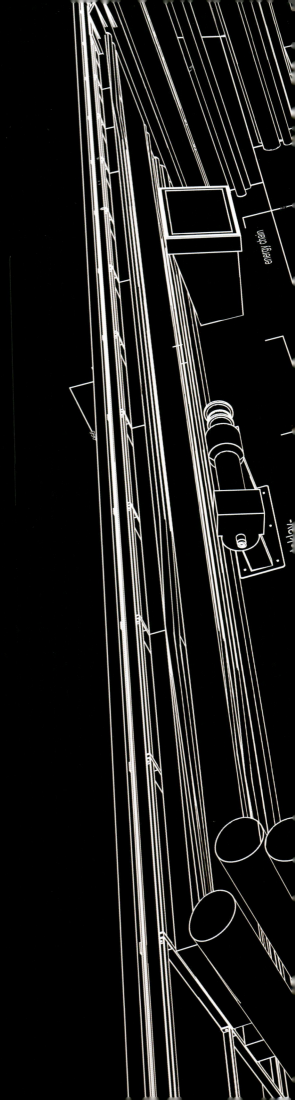

EINE AUSSTELLUNG ENTSTEHT
AN EXHIBITION COMES INTO BEING

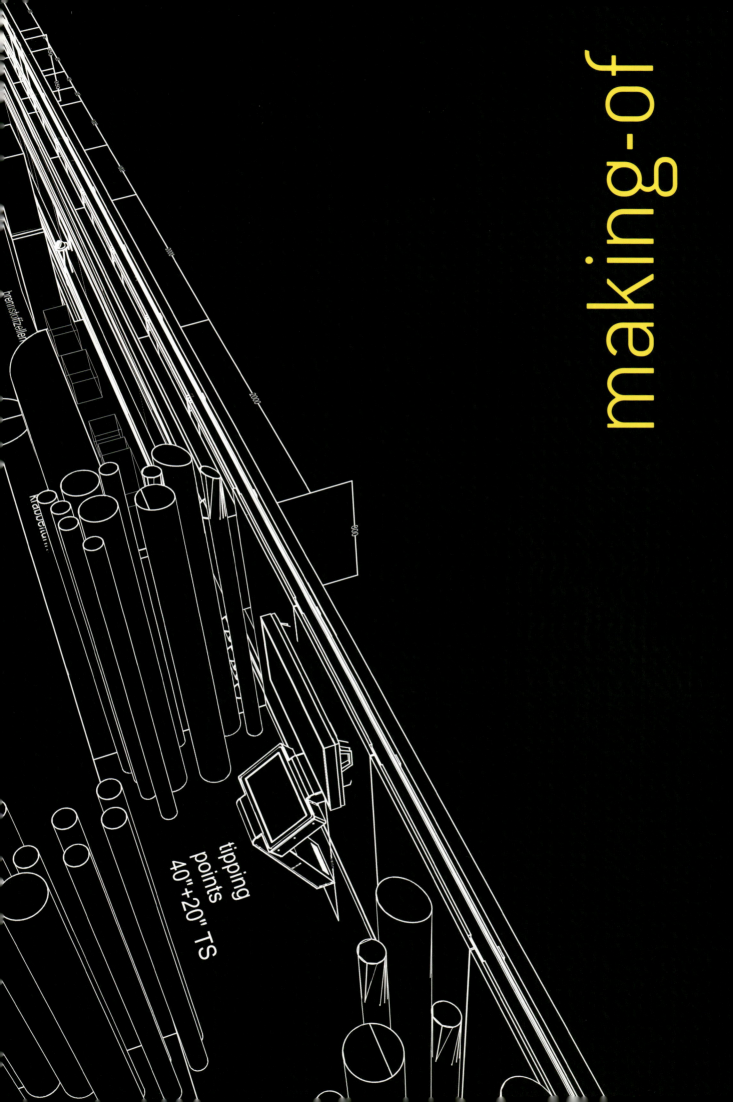

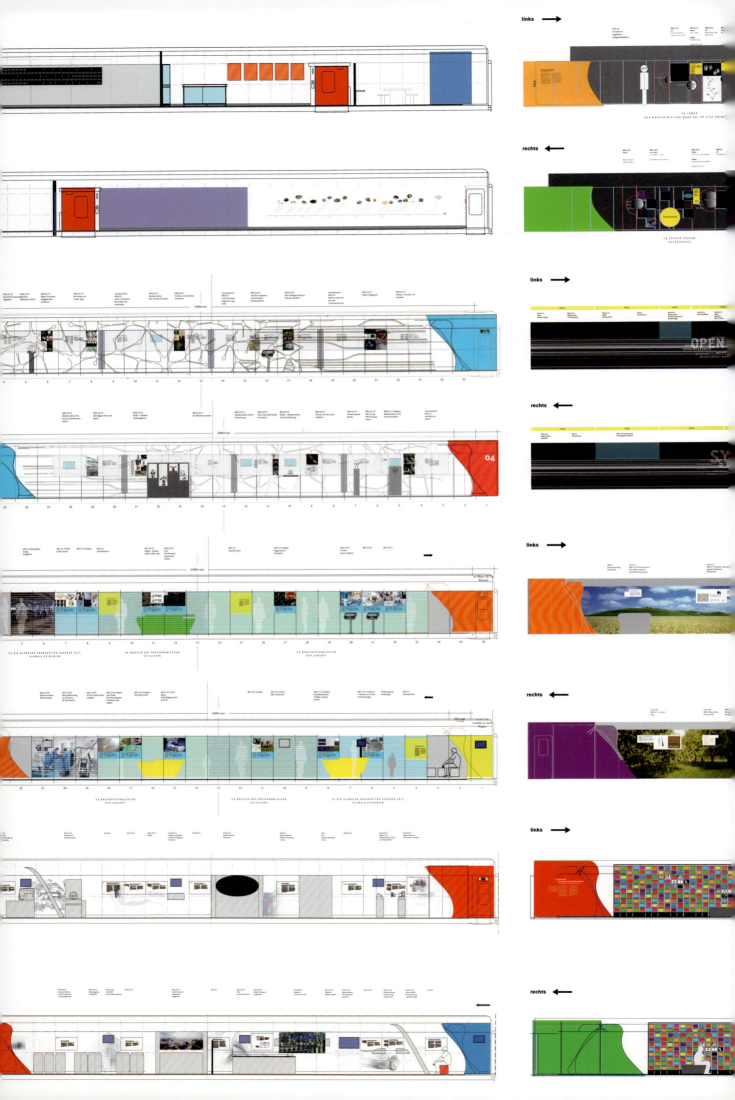

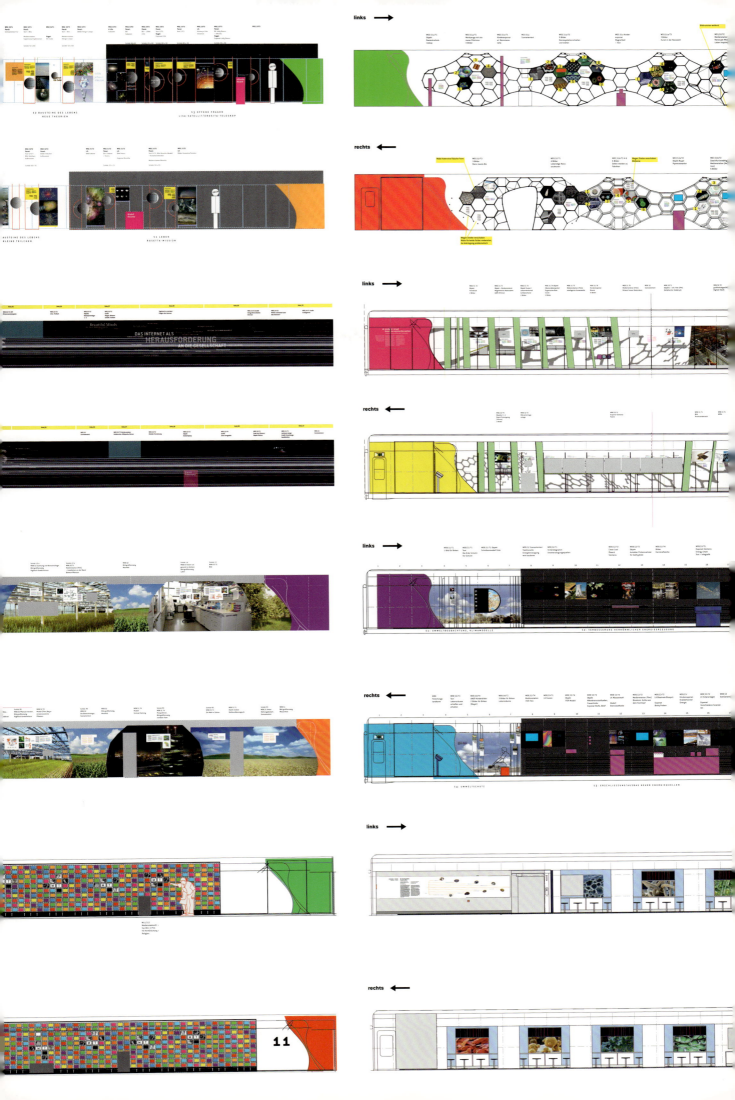

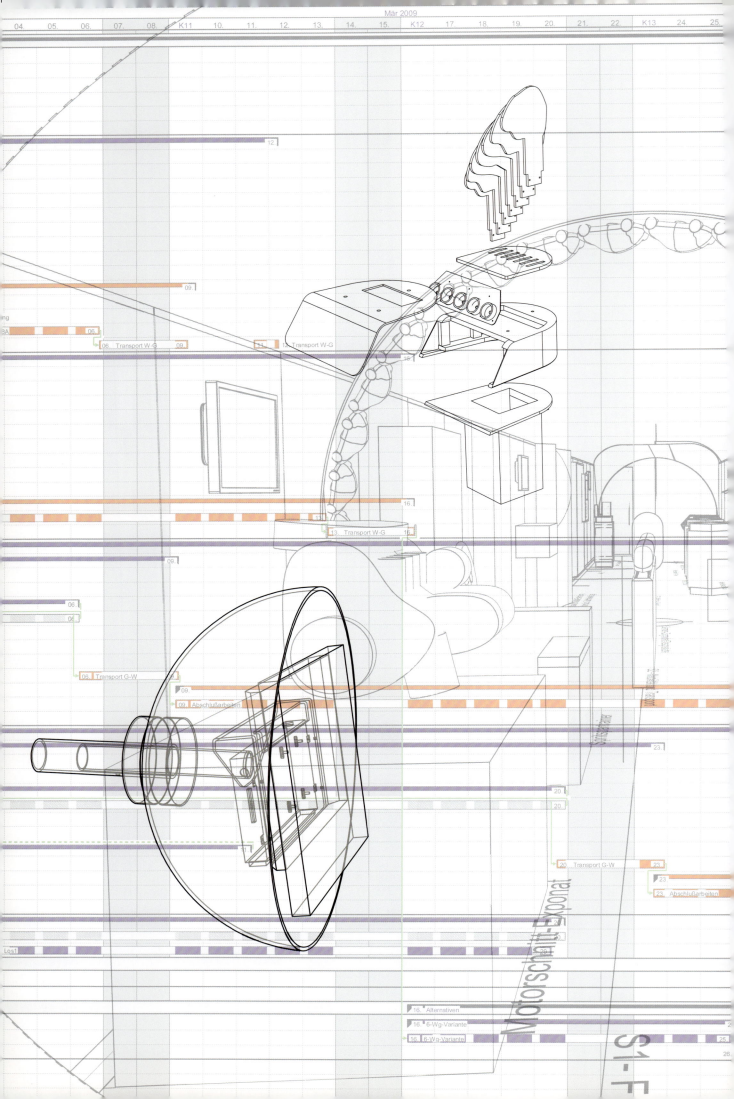

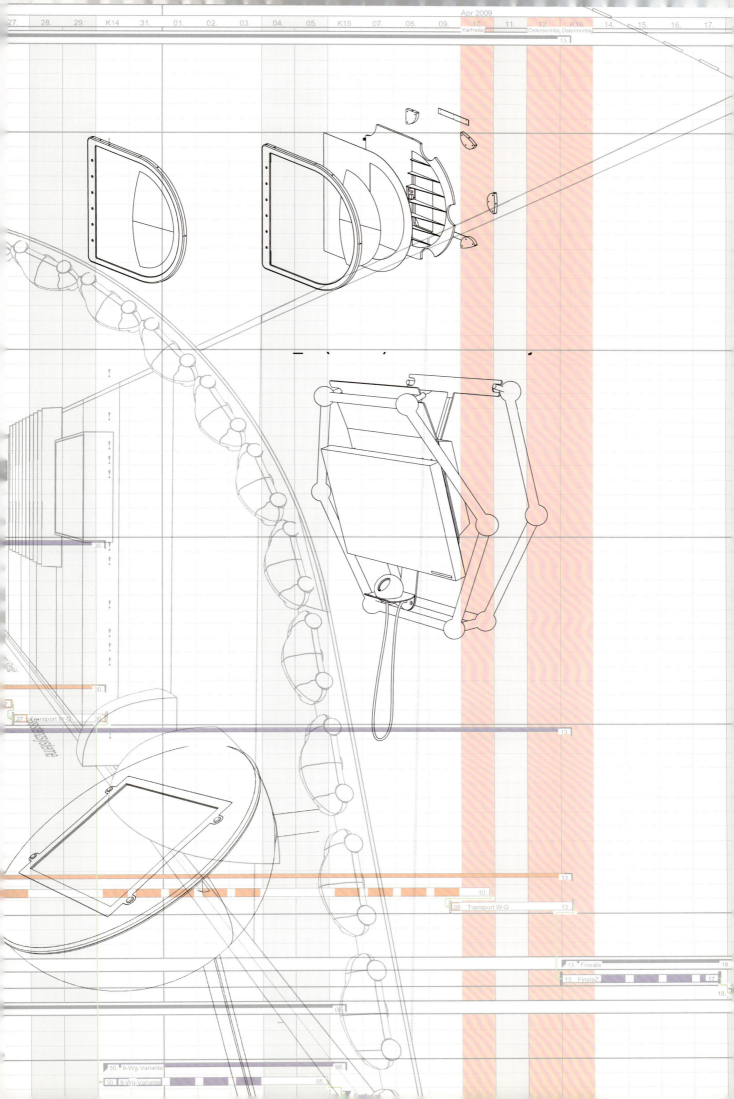

August 2008. Es geht los Während das Kuratorenteam beginnt, Inhalte zu recherchieren und Originalobjekte einzuwerben, entwickelt die Ausstellungsagentur das Design für die Wagen und erste interaktive Exponate. Noch ist eine der wichtigsten Fragen nicht gelöst: Wie können in kurzer Zeit zwölf Zugwagen beschafft werden?
Oktober 2008. Die Zeit läuft Es bleiben nur fünf Monate, um die zwölf bereits stillgelegten und schwer durch Korrosion beschädigten Reisezugwagen wieder verkehrstüchtig zu machen und in Ausstellungsräume zu verwandeln. Die von der Deutschen Bahn gelieferten Wagen stammen zum Teil aus den 1960-er Jahren.
Der erste Schritt Bahnmitarbeiter im Werk Wittenberge reparieren die Wagen und entkernen sie. Um Platz für die Ausstellungsszenografie zu schaffen, müssen sämtliche Inneneinbauten vom Gepäcknetz bis zur Wandverkleidung entfernt werden. **Bau einer Ausstellung** Das 65-köpfige Ausstellungsteam in Berlin installiert die Klimaanlage und Elektrik, baut Fußböden und Wände, die Ausstellungsszenografie und schließlich rund 150 Exponate und Medienstationen ein. Dafür sind in jedem der Wagen etwa 2.000 Arbeitsstunden notwendig.
Eine technische Herausforderung Die Stromversorgung der Ausstellung und die Klimatisierung der Wagen muss sich den besonderen bahntechnischen Anforde-

rungen stellen. Um Feuersicherheit zu gewährleisten, benötigen alle Einbauten eine TÜV-Prüfung. **Den Zeitplan einhalten** Während eines Kälteeinbruchs im Januar wird die Arbeit auch bei -15 °C nicht unterbrochen. Um mit kälteempfindlichen Materialien jetzt noch arbeiten zu können, müssen die Wagen mit Wärmeaggregaten geheizt werden. **Der Bio-Nano-Wagen entsteht** Metallstangen und Verbindungsknoten bilden das Grundgerüst der Ausstellungsarchitektur. In einer Werkstatt werden in Handarbeit organische anmutende Formen aus einem Spezialschaum modelliert, bevor die Bauteile in den Zug gelangen. **Ein Wald im Wagen** 50 Birkenstämme sind Teil einer Inszenierung, die das Thema Umwelt und Energie präsentiert. Trotz der achtwöchigen Trocknung beginnen einzelne Stämme auszuschlagen, als sie im Frühjahr in den Zug eingebaut werden. **Ein bedrucktes Gesamtkunstwerk** Alle Wagen und die Lokomotive erhalten eine grafisch gestaltete Außenhaut, die 1.300 m² umfasst. Im Inneren des Zuges gibt es mit den Wandgestaltungen, Ausstellungstexten und Fotos weitere 900 m² gedruckte Grafik.

August 2008. The journey begins While the team of curators starts to research the content and acquire original objects, the exhibition agency develops the exhibition design and the first interactive exhibits. One of the most important questions, however, is still open: How can twelve train carriages be obtained at short notice? **October 2008. Time is moving on** just five months to make the twelve disused and heavily corroded passenger train carriages roadworthy again and convert them into exhibition spaces. Some of the carriages which were supplied by Deutsche Bahn originate from the 1960s. **The first step** Railway employees at the Wittenberge rail factory repair and gut the carriages. All of the interior fittings in the carriages – from baggage nets to panelling – have to be removed to create space for the exhibition scenography. **Setting up an exhibition** The 65-member exhibition team in Berlin installs the air-conditioning and electrical system, constructs floors and walls, the exhibition scenography, and, finally, the approximately 150 exhibits and media stations. This requires around 2,000 hours of work in each of the carriages. **A technical challenge** The exhibition power supply and air-conditioning for the carriages have to meet special railway-engineering requirements. To guarantee fire safety, for example, all fixtures must undergo a safety standards inspection. **Keeping on schedule** Even during a major cold spell in January with temperatures of -15 °C, the work is not interrupted. The carriages must be warmed with heaters to enable the team to work with materials sensitive to cold. **The bio-nano carriage emerges** Metal rods and connection nodes form the basic framework for the exhibition architecture. Forms imitating organic shapes are modelled in a workshop by hand from special foam before the components reach the train. **A forest in a carriage** 50 birch trunks are part of a presentation about the environment and energy. Despite the eight weeks drying time, individual trunks start to sprout after they are installed in the train in spring. **An impressive ›total work of art‹** The graphic design is applied to the exterior of all of the carriages and locomotives, and covers a total of 1,300 m². Inside the train, the wall designs, exhibition texts and photographs combine to form a further 900 m² of printed graphics.

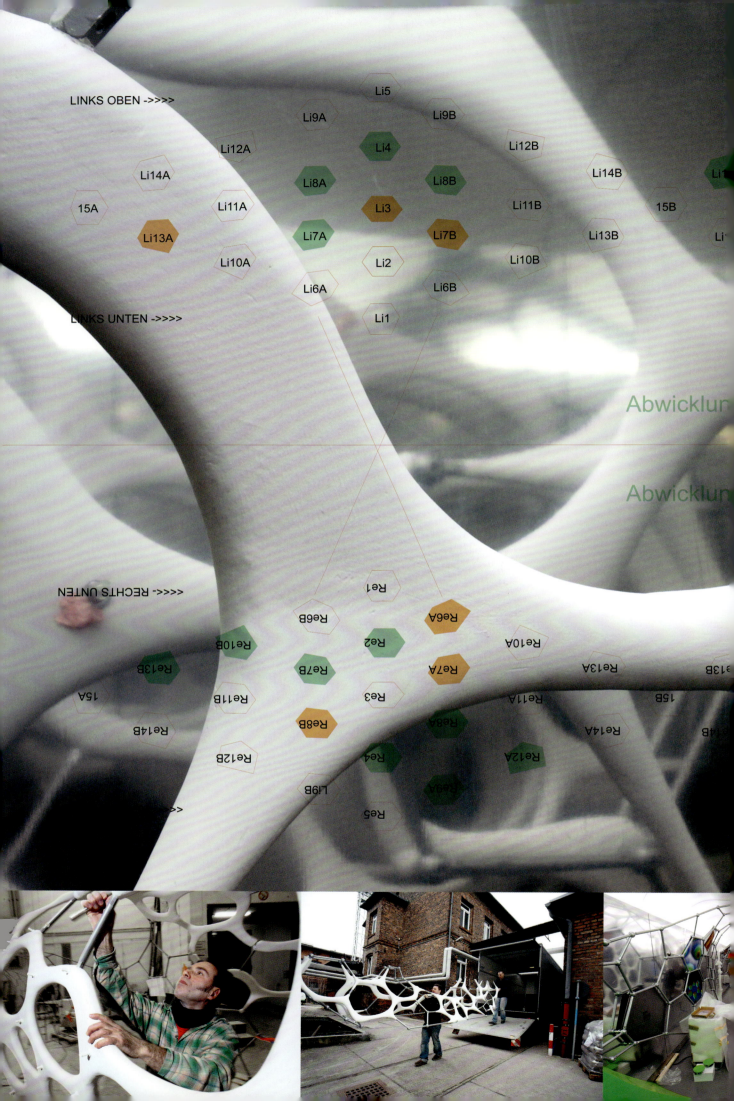

WAS KOMMT AUF UNS ZU?
WHAT ARE WE FACING?

expedition zukunft

Zukunft der Vergangenheit: Wissenschaft und Technik beeinflussen immer stärker unser Leben. Bewusst wird uns das oft erst im Rückblick. Konnten unsere Urgroßeltern eine Welt erahnen, die von weltumspannenden Computer-, Flug- und Kommunikationsnetzen geprägt wird? Wie weit reichen unsere Vorstellungen einer Welt in fünfzig Jahren? *Future of the past: Our lives are increasingly influenced by science and technology. Often we only realize this in retrospect. Could our great-grandfathers have imagined a world shaped by the global networks of computers, air connections and communication? What is our vision of the world in fifty years?*

Apfelzet

Was kommt auf uns zu?

Wir stehen am Beginn des 21. Jahrhunderts. Das vergangene Jahrhundert war geprägt von einer immer stärkeren Beschleunigung der wissenschaftlich-technischen Entwicklung, die zunehmend unsere Lebenswelten beeinflusst hat. Was wird uns die Zukunft bringen? Heute werden die Weichen dafür gestellt, wie wir künftig leben und arbeiten, wie alt wir werden, wie gesund wir bleiben oder welche Produkte und Leistungen wir nutzen können. All das hängt entscheidend davon ab, wie es Deutschland gelingt, sich einen führenden Platz in einer globalen Wissensgesellschaft zu sichern. Grundvoraussetzung dafür ist sowohl ein massives Engagement in die Produktion neuen Wissens und innovativer Technologien als auch in deren Anwendung und Verwertung.

What are we facing?

We are standing at the beginning of the 21st century. The past century was characterised by the ever-accelerating pace of scientific and technological development, which increasingly affects every aspect of our lives. What will the future bring? Today we set the course for how we will live and work in the future, how old we will become, how healthy we will remain or which products and services will be available to us. All of this depends on how well Germany succeeds in securing a leading position in the global knowledge society. The basic prerequisite for this is a massive commitment to both the production of new knowledge and innovative technologies and their application, and exploitation.

DIE SUCHE NACH DEN URSPRÜNGEN
THE SEARCH FOR OUR ORIGINS

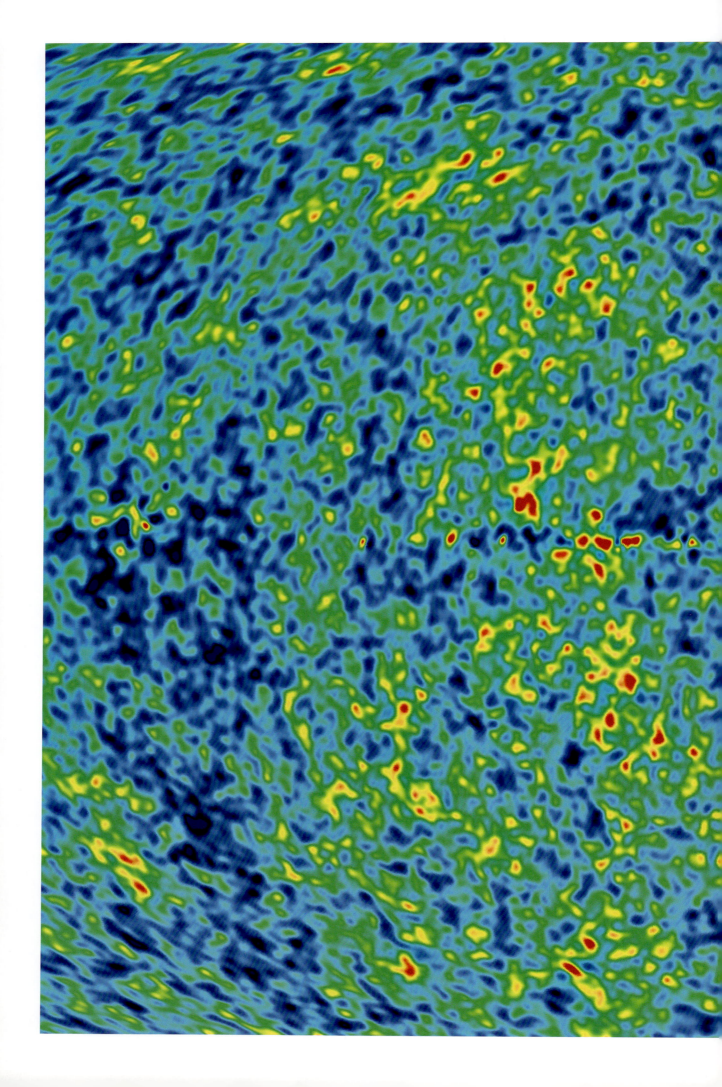

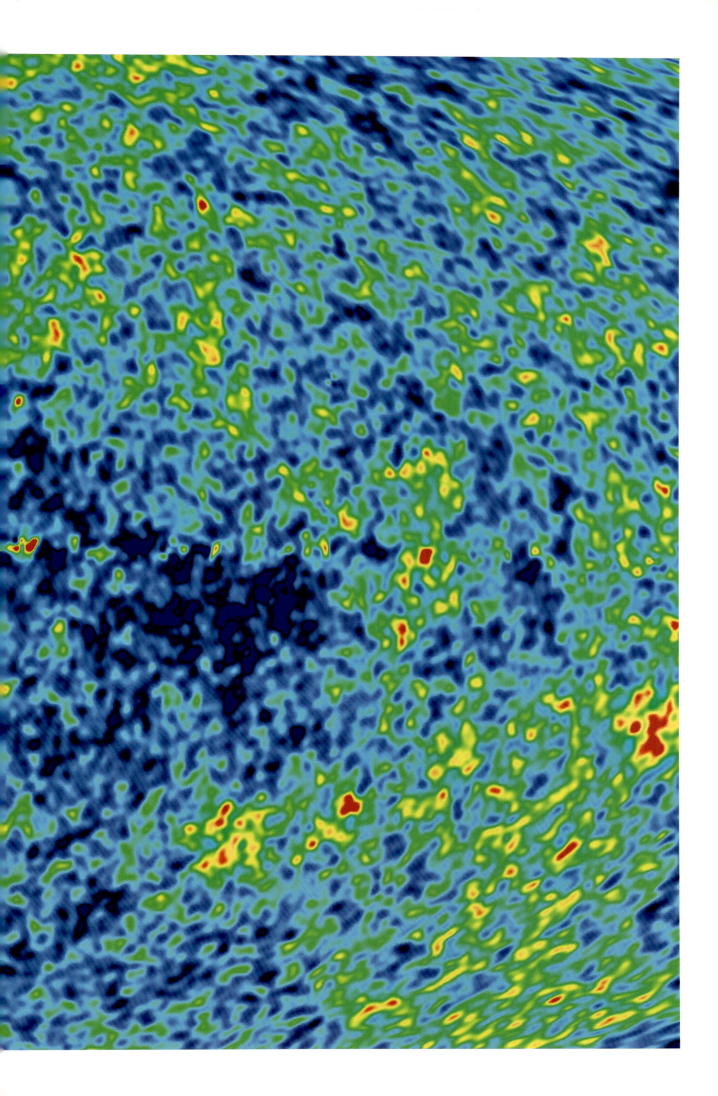

1

2

3

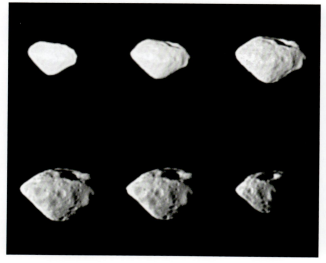

4

5

Die Suche nach den Ursprüngen

Das Universum scheint wie für uns gemacht. In einer Welt mit nur leicht veränderten Naturkonstanten wäre menschliches Leben vielleicht gar nicht entstanden. Doch es gibt uns, und wir können uns fragen: Woher kommen wir? Wohin gehen wir? Der Mensch steht am Ende einer langen Entwicklung. Die Evolution des Lebens auf der Erde ist dabei ein relativ junger Prozess. Davor entstanden im All über Jahrmilliarden jene Bausteine, aus denen alles besteht. Aus Beobachtungen und Experimenten kennen wir heute die Elementarteilchen und wissen welche Kräfte zwischen ihnen wirken. Doch das Universum besteht nur zu vier Prozent aus dieser ›gewöhnlichen‹ Materie. Physiker versuchen herauszufinden, woraus die restlichen 96 Prozent bestehen. **Leben auf der Erde – und anderswo?** *Der Mensch ist das einzige Wesen, das nach seiner Herkunft fragen kann. Inzwischen suchen wir nicht nur auf der Erde nach unseren Wurzeln, sondern auch im Weltraum und blicken dabei immer weiter in die Vergangenheit zurück. Das Leben auf der Erde begann vor 3,5 Milliarden Jahren mit den ersten primitiven Organismen. Diese konnten Stoffe aus ihrer Umgebung aufnehmen und sich fortpflanzen. Ihre Bestandteile, die ersten komplexen Moleküle, bildeten sich auf der Erde oder kamen mit Kometen auf unseren Planeten. Flüssiges Wasser gilt als unverzichtbare Voraussetzung für die Existenz von Leben, und das gibt es auch auf anderen Planeten oder Monden.* **Die Entwicklung des Menschen** Vor etwa 200.000 Jahren entwickelte sich der moderne Mensch und eroberte vor ungefähr 100.000 Jahren die Welt – wahrscheinlich von Afrika aus. Archäologische Funde zeichnen ein immer genaueres Bild der menschlichen Urgeschichte. Um die wertvollen Objekte zu schützen und sie gleichzeitig möglichst vielen Forschern zugänglich zu machen, werden sie gescannt und in Online-Bibliotheken veröffentlicht. **Ursuppe oder Meteoriten?** Vor etwa 3,5 Milliarden Jahren gab es auf unserem Planeten erstes mikrobielles Leben. Woraus dieses entstand – und ob es nur auf der Erde entstehen konnte – wissen wir bisher nicht. Einige Aminosäuren, die elementaren Bausteine des Lebens, können unter den richtigen Bedingungen aus wenigen anorganischen Zutaten entstehen. Ob dies auf der jungen Erde geschah, oder ob das erste Leben aus dem All kam, ist noch ungeklärt. **Wasser auf anderen Planeten?** Die Erde ist vielleicht doch nicht einzigartig. Die Raumsonde Phoenix entdeckte 2008 Eis auf dem Mars; vielleicht gab es auch frühe Lebensformen auf dem roten Planeten. Inzwischen sind mehr als 330 Planeten außerhalb unseres Sonnensystems bekannt, darunter einige, auf denen es flüssiges Wasser geben könnte. Auch Planeten und Monde in unserem Sonnensystem sind vielleicht nicht so lebensfeindlich wie gedacht. **Wir sind aus Sternenstaub** *Menschen, Tiere, Pflanzen, Erde, Luft – alles besteht aus der ›Asche‹ längst erloschener Sterne. Durch astronomische Beobachtungen, Modellrechnungen und raffinierte Experimente auf der Erde kennen wir heute die wichtigsten Bausteine der Materie. Einige leichte Elemente, wie Wasserstoff und Helium, entstanden bereits beim Urknall vor mehr als 13 Milliarden Jahren. Die schweren Elemente, etwa Kohlenstoff und Sauerstoff, bildeten sich dann in Sternen und Supernovae, dem explosiven Ende massiver Sterne. Einige dieser extremen Bedingungen können Physiker in Experimenten auf der Erde nachstellen und so ins Innerste der Materie blicken. Und vielleicht entstehen dabei auch ganz neue Teilchen, die bisher nur theoretisch vorhergesagt werden konnten.* **Die Entstehung der Elemente** Etwa drei Minuten nach dem Urknall hatte sich das Universum so weit abgekühlt, dass sich leichte Elemente aus Protonen und Neutronen bilden konnten. Als die ersten Sterne zu leuchten begannen, entstanden in ihrem Innern schwerere Elemente durch Fusionsreaktionen. Für die schwersten Elemente waren aber noch höhere Energien nötig, wie sie bei Supernova-Explosionen herrschen. **Viel Energie für kleinste Teilchen** Fast wie beim Urknall, stoßen in riesigen Beschleuniger-Röhren Teilchen mit nahezu Lichtgeschwindigkeit zusammen und produzieren dabei neue Partikel. Durch

---

Vorige Doppelseite: Kosmische Mikrowellenhintergrundstrahlung. *Preceding spread: Cosmic microwave background radiation.* ——— 1 Der Komet McNaught über dem Pazifik 2007. *Comet McNaught over the Pacific in 2007.* ——— 2 Temperaturmessung an einem Schwarzen Raucher: Mit bis zu 400 °C heißem Wasser ähneln diese möglicherweise den extremen Umweltbedingungen in der frühen Erdgeschichte. *Measuring the temperature of a black smoker. With hot water of up to 400 °C, these phenomena might resemble the extreme environmental conditions that prevailed when the Earth was very young.* ——— 3 Fossilien sind selten und zerbrechlich. Forscher benutzen deshalb 3D-Modelle im Computer und als 3D-Druck, wie diesen Neandertaler-Schädel. *Fossils are rare and fragile. Scientists therefore use 3D models on computers and as 3D print outs, like this Neanderthal skull.* ——— 4 Bilder des Asteroiden Šteins, aufgenommen beim Vorbeiflug des Rosetta-Satelliten. *Images of the Šteins asteroid taken from the Rosetta satellite as it flew past.* ——— 5 Falschfarben-Aufnahme des Saturnmondes Titan während des Vorbeiflugs der Raumsonde Cassini. *Pseudo-colour image of Titan, one of Saturn's moons, taken during a fly-by of the Cassini probe.*

Vorige Doppelseite *Preceding spread* NASA/WMAP 1 Sebastian Deiries, ESO 2 MARUM, Universität Bremen 3 Max-Planck-Institut für evolutionäre Anthropologie, Leipzig 4 ESA, 2008 MPS for OSIRIS Team/UPD/LAM/IAA/RSSD/INTA/UPM/DASP/IDA 5 NASA/JPL/Space Science Institute

solche Beschleuniger-Experimente kennen wir heute die kleinsten Bausteine der Materie. Doch das so genannte Standardmodell der Teilchenphysik sagt noch weitere Partikel voraus, wie etwa das Higgs-Teilchen, das für die Masse verantwortlich ist. ——— **Von Saiten und Schleifen** Teilchen gleichen winzigen Kugeln, oder? Diese Vorstellung ist nicht ganz richtig, wenn es noch weitere Raumdimensionen gibt, die auf kleinstem Raum gefaltet sind. Laut der Stringtheorie gibt es winzige, eindimensionale Saiten in einem zehndimensionalen Raum, die auf viele unterschiedliche Arten schwingen können. Elementarteilchen sind damit lediglich Zustände ein und derselben Sorte fundamentaler ›Strings‹. ——— **96 Prozent des Universums liegen im Dunkeln** *Die bekannte Materie macht nur vier Prozent des Universums aus. Knapp ein Viertel ist Dunkle Materie und der Rest Dunkle Energie. Doch auch wenn wir die vielen Formen der normalen Materie gut kennen, geben uns die anderen Komponenten noch Rätsel auf. Aus astronomischen Beobachtungen können Kosmologen inzwischen bestimmte Parameter unseres Universums sehr genau bestimmen. Effekte, die auf der gravitativen Anziehungskraft von Sternen und Galaxien beruhen, zeigen uns, dass es ein vielfaches der Materie geben muss, die man sehen kann. Die beschleunigte Expansion unseres Universums lässt sich nur durch eine ›Dunkle Energie‹ erklären, über deren genaue Natur aber derzeit nur spekuliert werden kann.* ——— **Wir kennen nur vier Prozent** Sterne und Planeten bestehen aus den vier Prozent normaler Materie. Direkt sichtbar sind aber nur Sterne und heißes Gas, die selbst leuchten. Beobachtungen bei unterschiedlichen Wellenlängen geben Aufschluss über verschiedene physikalische Eigenschaften. Da das Licht der beobachteten Objekte oft Jahrmilliarden zu uns unterwegs ist, kann man damit auch in die Vergangenheit blicken. ——— **Woraus besteht das Universum?** Die bekannte Materie reicht nicht aus, um bestimmte astronomische Beobachtungen im Universum zu erklären. Wertet man Rotationskurven von Sternen und heißem Gas in Galaxien aus, so zeigt sich, dass in Galaxien weit mehr Materie vorhanden sein muss, als wir sehen können. Diese Vermutung wird durch die Beobachtung von Gravitationslinsen bestätigt. ——— **Dunkle Materie und Dunkle Energie** Auch wenn man sie nicht sehen kann, so wissen wir heute schon einiges über die Dunkle Materie, etwa dass sich ihre Teilchen viel langsamer als mit Lichtgeschwindigkeit bewegen. Bisher wurde noch kein Teilchen der Dunklen Materie beobachtet, sie treten also nur schwach mit anderen Teilchen in Wechselwirkung. Für die Dunkle Energie gibt es noch keine befriedigende Erklärung, die zu den astronomischen Beobachtungen passen würde.

Hintergrund: Im Sternhaufen NGC 2467 entstehen gegenwärtig viele neue Sterne aus interstellarem Staub und Gas. Er befindet sich in einer Distanz von mehr als 25.000 Lichtjahren. *Background: Many new stars are currently born from interstellar gas and dust in the stellar cluster NGC 2467. Its distance is more than 25,000 light years.* ——— 1–3 Nach Fertigstellung wird das ›IceCube‹ Neutrino-Observatorium aus mehr als 5.000 digitalen, optischen Modulen in dutzenden Bohrlöchern bestehen, verteilt über einen Kubikkilometer im antarktischen Eis. *When completed, the ›IceCube‹ detector will consist of more than 5,000 digital optical modules in dozens of boreholes, spanning a cubic kilometre of deep Antarctic ice.*

Hintergrund *Background* ESO/Max-Planck-Institut für Astronomie, Heidelberg 1–3 IceCube/NSF

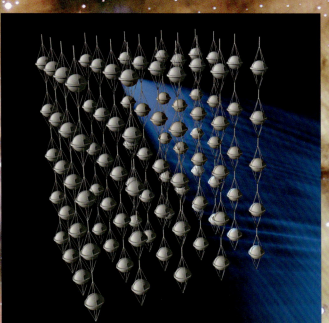

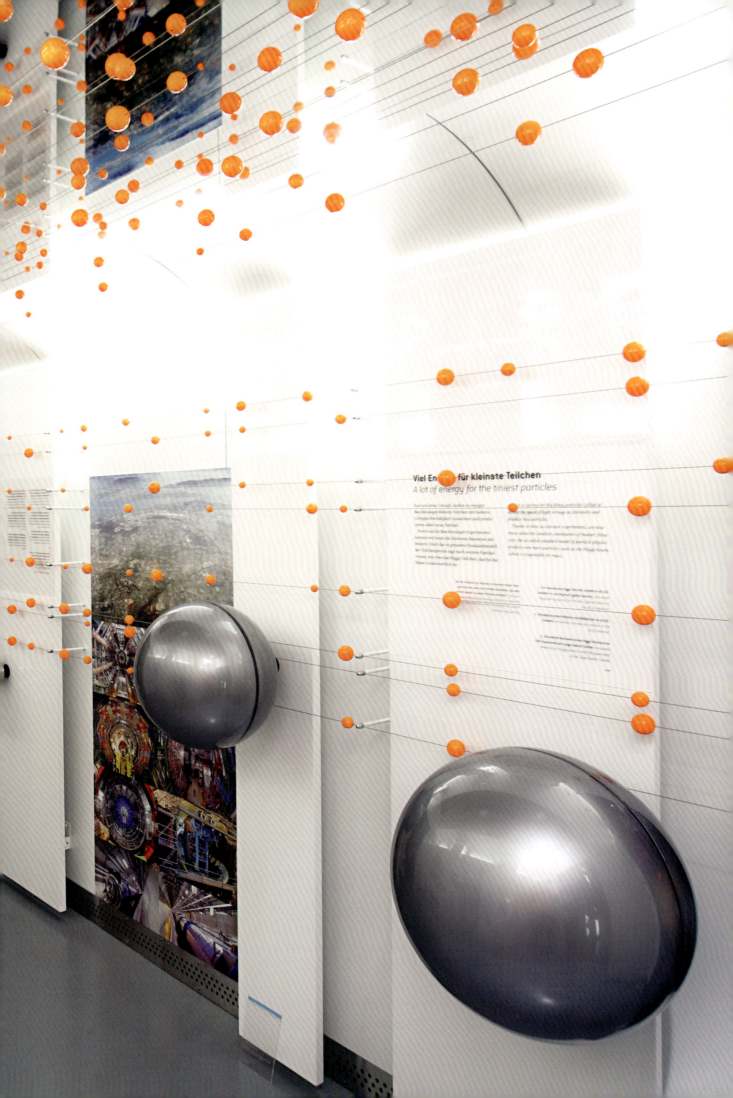

The search for our origins

The Universe seems to have been made just for us. If the physical constants were even just slightly altered, we might never have seen the emergence of human life. However, we are indeed here and we might ask ourselves: Where did we come from? Where are we going? A long process led to humankind. The evolution of life on Earth is thereby a relatively recent development. Before that, over billions of years in the universe, the elements were formed from which everything is made. Observations and experiments have taught us about elementary particles and the forces that act between them. However, this ›normal‹ matter makes up only four percent of the universe. Physicists are trying to find out what the remaining 96 percent consist of.

Life on Earth – and elsewhere? *Humans are the only beings that can question their origins. By now, we are searching for our roots not only on Earth but in space as well. This search is leading us to look further and further into our past. Life on Earth began 3.5 billion years ago with the first primitive organisms. These organisms were able to absorb substances from their surroundings and to reproduce. Their constituents, the first complex molecules, either formed on Earth or they arrived on our planet with comets. Liquid water is an essential prerequisite for the existence of life; and this exists also on other planets and moons.*

Human development Modern humans developed about 200,000 years ago and spread throughout the world some 100,000 years ago – probably from Africa. Archaeological finds are delivering an increasingly detailed picture of human prehistory. In order to protect these valuable objects and to make them accessible to as many researchers as possible, they are scanned and published in online-libraries.

Primeval soup or meteorites? Microbial life first emerged on our planet around 3.5 billion years ago. We do not yet know what it developed from, nor do we know whether it could only have arisen on Earth. Under the right conditions, some amino acids, the elementary components of life, can be created from a few inorganic ingredients. It is still not known whether this is what happened on the young Earth or whether life first came from space.

Water on other planets? The Earth might not be unique after all. In 2008, the Phoenix spacecraft discovered ice on Mars. Were there early life forms on the

1 Computeranimationen wie diese visualisieren die Quanten-Geometrie, wie sie aus der Theorie der Schleifen-Quantengravitation folgt. *Computer animations like this one visualize quantum geometry as predicted by the theory of loop quantum gravity.* 2 Simulierter Nachweis eines Higgs-Teilchens im CMS-Experiment am Large Hadron Collider. *Simulated detection of a Higgs boson in the CMS experiment at the Large Hadron Collider.* 3 Das ATLAS-Experiment sucht nach dem Verursacher der Masse, weiteren Raumdimensionen, mikroskopischen schwarzen Löchern und Kandidaten für die Dunkle Materie. *The ATLAS experiment will look for the origin of mass, extra dimensions of space, microscopic black holes, and evidence for dark matter candidates.* 4 Das CMS-Experiment erforscht die Eigenschaften extrem heißer und dichter Materie und versucht das Higgs-Teilchen nachzuweisen. *The CMS experiment will investigate the properties of extremely hot, dense matter and try to detect new particles such as the Higgs boson.* 5 Das LHCb-Experiment untersucht, was direkt nach dem Urknall passierte und warum heute keine Antimaterie, sondern nur noch Materie übriggeblieben ist. *LHCb is an experiment set up to explore what happened immediately after the Big Bang and why no antimatter but only matter remains today.* 6 Der Schwerionen-Detektor ALICE wird Materie bei extrem hoher Energiedichte untersuchen, wo ein neuer Materiezustand, das Quark-Gluon-Plasma, erwartet wird. *The ALICE heavy-ion detector will study matter with extremely high energy densities, where the formation of a new phase of matter, the quark-gluon plasma, is expected.*

1 Max-Planck-Institut für Gravitationsphysik, Potsdam 2–6 CERN

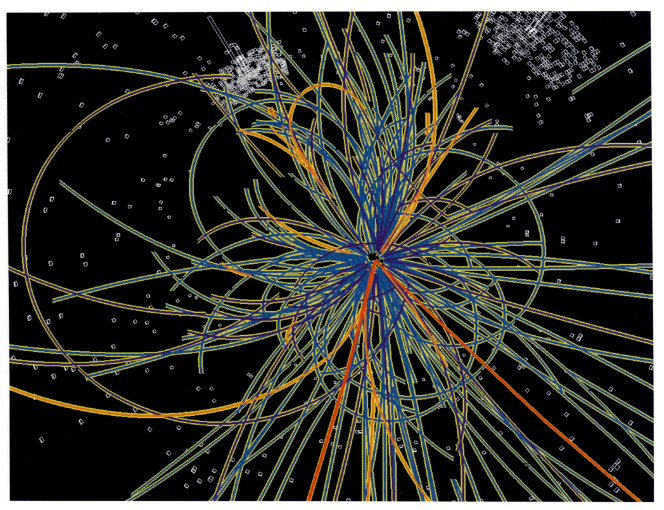
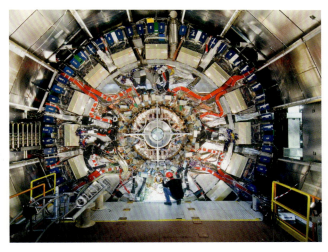
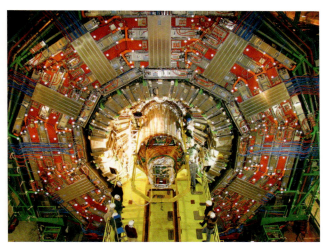
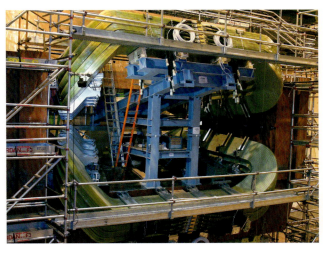
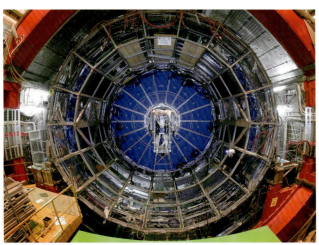

Red Planet as well? More than 330 planets are now known outside of our solar system, and there could be liquid water on some of them. Furthermore, some planets and moons in our solar system are not as hostile to life as was previously thought. **We are stardust** *Humans, animals, plants, earth and air – are all made of the ›ash‹ of stars that became extinct a long time ago. Through astronomical observations, mathematical models and sophisticated experiments we now know what the most important components of matter are. Some light elements, such as hydrogen and helium, were created during the Big Bang more than 13 billion years ago. Then heavy elements like carbon and oxygen formed in stars and in supernovae, the explosions that represent the death of massive stars. Physicists can recreate some of these extreme conditions in experiments on Earth and thus gain insight into the secrets of matter. This might also give rise to new particles that so far were predicted only in theory.* **The origin of the elements** Around three minutes after the Big Bang, the universe had cooled enough to allow light elements to form from protons and neutrons. When the first stars began to shine, heavier elements were formed in fusion reactions at their core. For the creation of the heaviest elements conditions of extremely high energy are necessary, such as those prevailing during supernova explosions. **A lot of energy for the tiniest particles** Almost as during the Big Bang, particles collide at almost the speed of light in huge accelerators and produce new particles. Thanks to these accelerator experiments, we now know about the smallest constituents of matter. However, the so-called standard model of particle physics predicts even more particles, such as the Higgs boson, which is responsible for mass. **Strings and loops** *Particles look like tiny spheres, right? This notion is not quite correct if there are additional dimensions folded into a very tiny space. According to string theory, there are tiny one-dimensional strings in a ten-dimensional space, which can vibrate in many different ways. Elementary particles are different states of one and the same variety of fundamental strings.* **96 percent of the universe is in the dark** *Known matter makes up only four percent of the universe. Just under a quarter is dark matter and the remainder is dark energy. Even though we are very familiar with the many forms of normal matter, the other components remain a puzzle. Astronomical observations allow cosmologists to define some parameters of our universe very precisely. Effects based on the gravitational attraction of stars and galaxies show us that there must be many times more matter than can be seen. The accelerating expansion of our universe can only be explained by ›dark energy‹ – but at the moment, it is only possible to speculate about its exact nature.* **We only know four percent** Stars and planets are made up of the four percent of normal matter. However, only stars and hot gases, which emit light, are directly visible. Observations at different wavelengths reveal information about different physical properties. As the light of observed objects has often travelled for billions of years before it reaches us, we thus can look into the past. **What does the universe consist of?** The matter that we know is not sufficient to explain certain astronomical observations in the universe. Analysis of the rotational curves of stars and hot gas in galaxies shows that they must contain much more matter than we can see. This assumption is confirmed by the observation of gravitational lenses. **Dark matter and dark energy** Even though we can't see it, we already know something about dark matter: for example, that its particles move much more slowly than at the speed of light. Up to now, there have been no observations of dark matter particles – showing that they interact only weakly with other particles. As yet there is no satisfactory explanation for dark energy that would be consistent with astronomical observations.

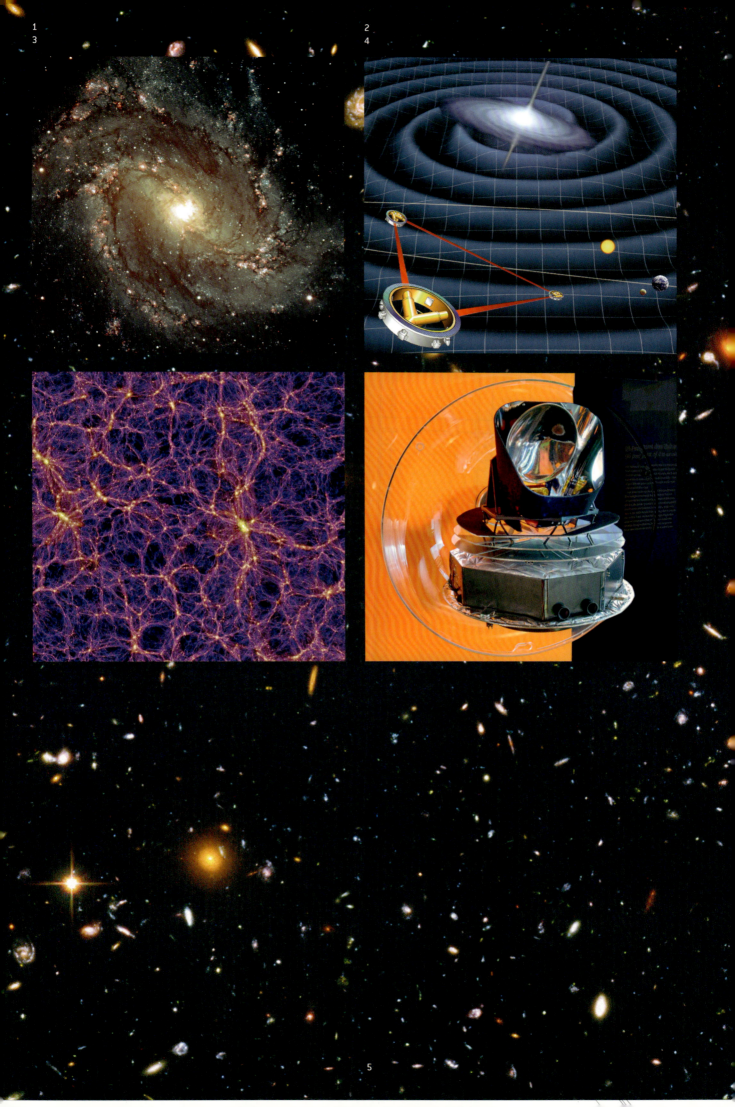

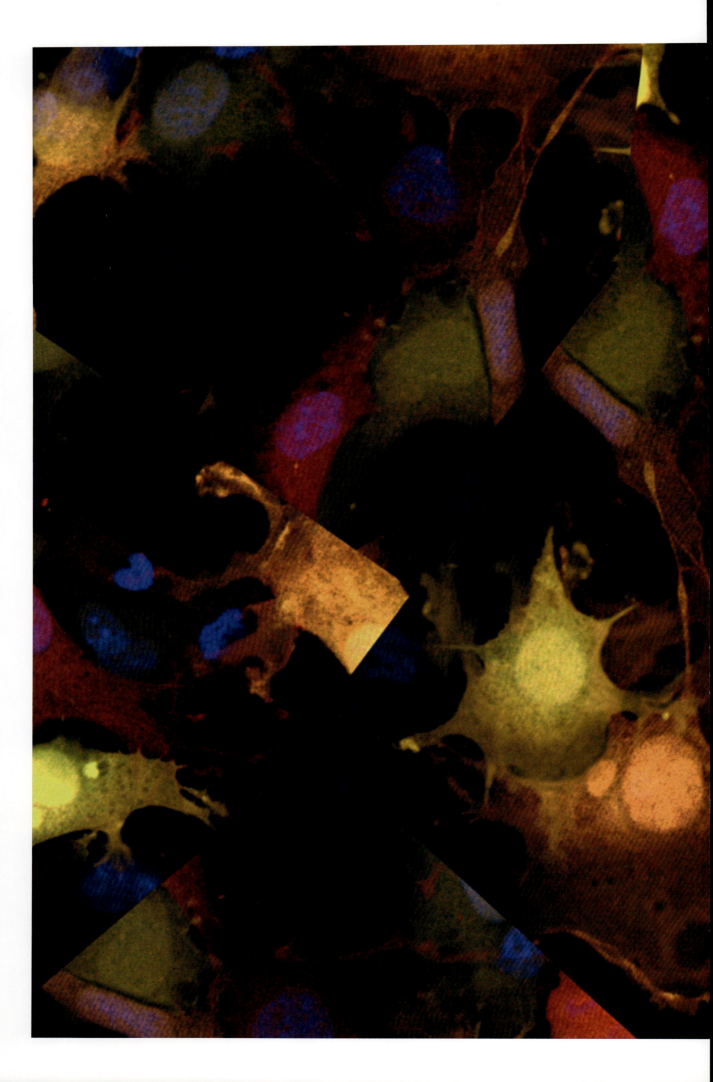

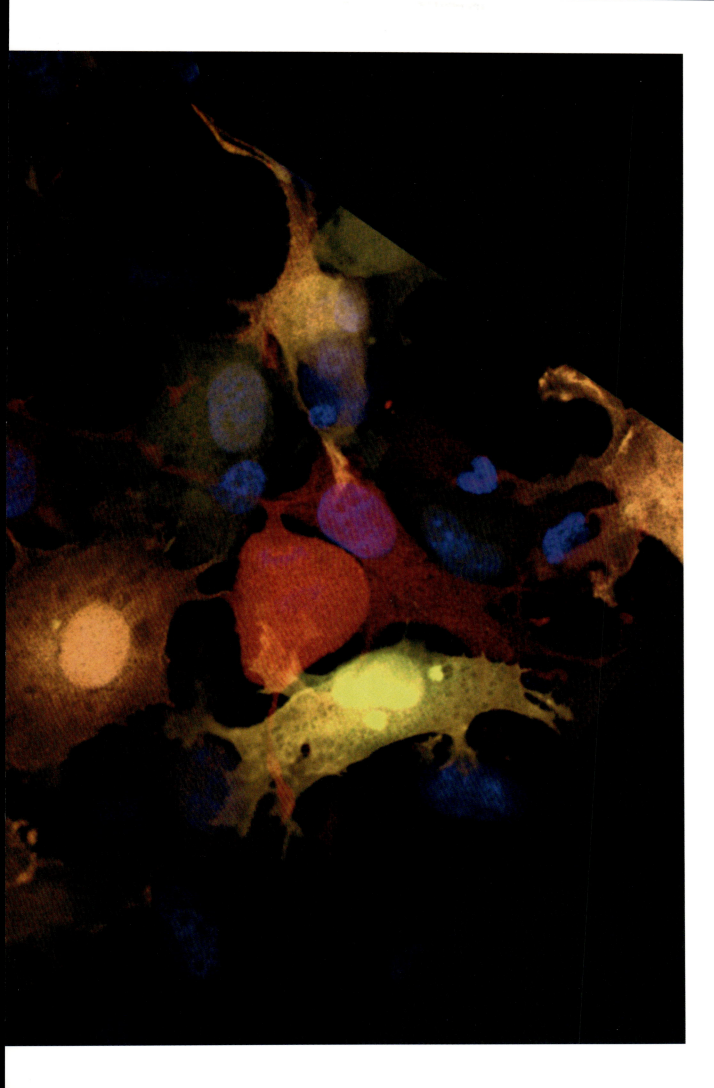

1

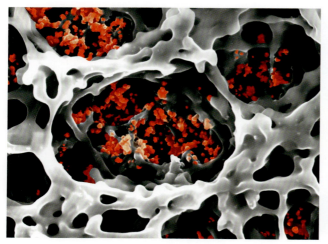

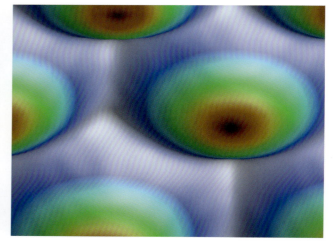

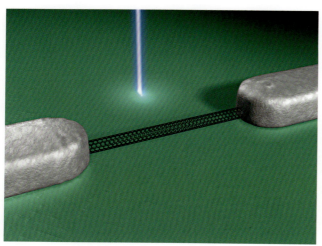

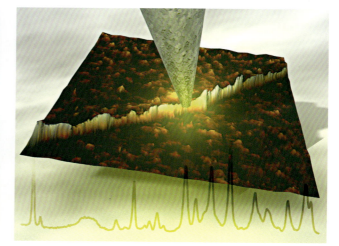

2 3
4 5

Nano- und Biowissenschaften verschmelzen

Der Mensch versteht heute viele physikalische, chemische und biologische Prozesse bis hin zur atomaren Ebene. Er wird fähig, tote und lebende Materie auf dieser winzigen Skala zu manipulieren. Die Grenzen zwischen Bio und Nano verschwimmen. Forscher entwickeln atomar präzise Fertigungsverfahren. Mit immer raffinierteren und aufwändigeren Methoden entschlüsseln sie Syntax und Semantik des Lebens – die Sprache der Gene. Integrierte Schaltkreise beginnen mit lebenden Zellen zu kommunizieren und Daten an Computer weiterzuleiten. Mikroorganismen entstehen, deren Verhalten und Stoffwechsel programmierbar sind: Die Konvergenz zwischen Nano- und Biowissenschaften schreitet unübersehbar voran. Die Synthetische Biologie ist die neue Nanotechnologie._____ *›Ganz unten ist eine Menge Platz‹ Der Physiker Richard P. Feynman hatte schon vor 50 Jahren die Vision vom ›Sehen‹ und Manipulieren einzelner Atome. Im Verlauf der vergangenen drei Jahrzehnte ist sein Traum wahr geworden, denn in der Nanowelt gelten andere Gesetze als in der Makrowelt. Die Nanowissenschaften ermöglichen eine extreme Miniaturisierung der Technik. Komplizierte Systeme werden immer kleiner, doch noch gibt es viel Spielraum nach unten. Die winzigen Baueinheiten der Nanowelt sind aber immer noch wesentlich größer als einzelne Atome oder Moleküle. Noch sind wir nicht in der kleinsten aller Welten angekommen. Wann können Forscher einzelne Atome und Moleküle perfekt und nach Bedarf zusammensetzen?*_____ **Werkzeuge mit atomarer Präzision** Die Verfahren, Materie im Nanobereich zu manipulieren, haben sich rasant entwickelt. Sie führten zur Geburt der Nanowissenschaften. Neben dem Abtasten von Oberflächen und der Vermessung atomarer Kräfte können Forscher schon heute mit dem Rasterkraftmikroskop einzelne Atome gezielt bewegen. Bis zum Bau von funktionstüchtigen Nanomaschinen ist es aber noch ein langer Weg._____ **Nanosysteme schalten und walten** Auch in der Nanowelt gibt es Werkzeuge: Motoren, Rotoren, Scheren, Gelenke. Mit ihren Pendants aus der Makrowelt haben sie wenig gemein. Bisher dienen die winzigen Nanogeräte zumeist dazu, extrem kleine Entfernungen oder extrem schwache Kräfte zu messen. In einigen Jahren könnten wir sie bereits in Sensoren, Solarzellen, elektronischem Papier oder biegsamen Bildschirmen finden._____ **Kunst in der Nanowelt**_____ Phantasie brauchen wir in der Wissenschaft besonders. Sie ist nicht nur Mathematik, nicht nur Logik, sondern ein wenig auch Schönheit und Poesie. *Maria Mitchell (1818–1889), amerikanische Astronomin*_____ Die Untersuchung von Nanostrukturen liefert fantastische Aufnahmen, die uns in bezaubernde Welten voller Geheimnisse versetzen. Von mystischen Grotten bis zu magischen Wäldern reichen die Aufnahmen aus dem Nanokosmos. Die Motive ähneln auf geheimnisvolle Weise Dingen der Makrowelt. Forscherinnen der Nanotechnologie beweisen hier ihr künstlerisches Auge._____ **Was folgt auf die Entschlüsselung des Genoms?** *Der Bauplan ganzer Genome höherer Organismen ist entschlüsselt. Die molekulare Beschreibung von lebenden Systemen wird immer umfassender. Einzelne Prozesse können bis ins Detail erklärt werden. Dennoch wissen wir nicht, wie Leben funktioniert. Eine rasante technologische Entwicklung erlaubt es, das gesamte molekulare Inventar von Zellen zu katalogisieren. Die Datenflut stellt Biologen vor Aufgaben, die nur noch gemeinsam mit Informatikern und Mathematikern gelöst werden können. Die Systembiologie soll von der Syntax zur Semantik des Lebens vordringen. Wie arbeiten Millionen von Molekülen zusammen, um Leben zu schaffen? Forscher wollen das Verhalten lebender Systeme vorhersagen und Mikroorganismen mit neuen Eigenschaften bauen.*_____ **Blick ins Innerste der Zelle** Mit hochauflösenden Methoden kann das Zellinnere beobachtet werden. Dadurch verstehen wir immer besser die Funktion und Wechselwirkung der molekularen Bausteine. Auch die Struktur unseres Erbguts, der DNS, ist kein Geheimnis mehr. Das Genom ist entschlüsselt und gibt stückweise seine Informationen preis. Künftig kennt jeder seine individuelle DNS-Sequenz und deren Bedeutung für sein Leben.

*Vorige Doppelseite: Lebende, angefärbte Zellen – Intravitalmikroskopie macht es möglich. Preceding spread: Living, stained cells – thanks to intravital microscopy.*_____ **1** *Siegeszug der Natur: Elektronenmikroskopische Aufnahme von der amorphen Schicht eines organischen Polymerfilms, der mit Kristallen durchsetzt ist. Triumph of Nature: Electron-microscopy image of an amorphous, organic polymeric layer which is interspersed with crystals.*_____ **2** *Durch eine neuartige Synthese stellen Forscher schnell und einfach leuchtende oder elektrisch leitende, zehn bis fünfzehn Nanometer große Kristalle her. A new kind of synthesis allows scientists to create glowing or electrically conductive 10–15 nanometer crystals quickly and easily.*_____ **3** *Die Kräfte zur Manipulation auf atomarer Ebene sind noch nicht vermessen worden. The forces required to manipulate atoms have not yet been measured. This image shows the tip-adsorbate energy landscape during manipulation of a cobalt atom in a scanning force microscope.* Dieses Bild zeigt die Energie-Landschaft zwischen der Spitze eines Rasterkraftmikroskops und einem Kobaltatom._____ **4** *Winzige Schalter: Mit einem Elektronenstrahl lässt sich die Leitfähigkeit einer Kohlenstoff-Nanoröhre lokal auf ein Tausendstel des ursprünglichen Werts verringern. Tiny switches: A beam of electrons can reduce the conductivity of a carbon nano tube locally to one-thousandth of its initial value.*_____ **5** *Die DNS wie einen Strichcode scannen: Eine Kombination von Raman-Spektroskopie und Rasterkraftmikroskopie könnte die Abfolge der Basen künftig direkt bestimmen. Scanning DNA like a barcode: a combination of Raman spectroscopy and scanning force microscopy could determine the gene sequence directly in the future.*

Vorige Doppelseite Preceding spread Friedemann Kiefer, Max-Planck-Institut für molekulare Biomedizin, Münster **1** Dr. Nina Rehmann, Arbeitsgruppe Prof. Meerholz, Universität zu Köln **2** DFG-Zentrum für Funktionelle Nanostrukturen der Universität Karlsruhe (TH) **3** Max-Planck-Institut für Mikrostrukturphysik, Halle/Saale **4** Christoph W. Marquardt & Dr. Ralph Krupke, Institut für Nanotechnologie, Forschungszentrum Karlsruhe **5** ISAS – Institute for Analytical Sciences, Dortmund, mit freundlicher Genehmigung von Volker Deckert

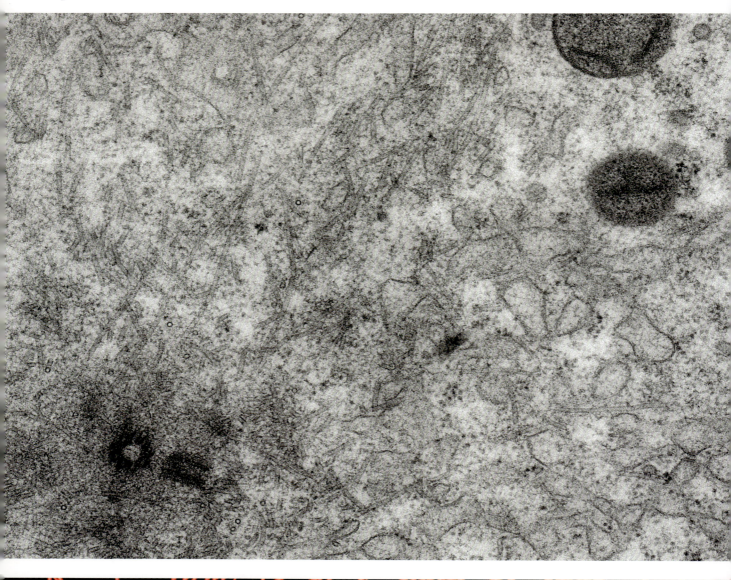

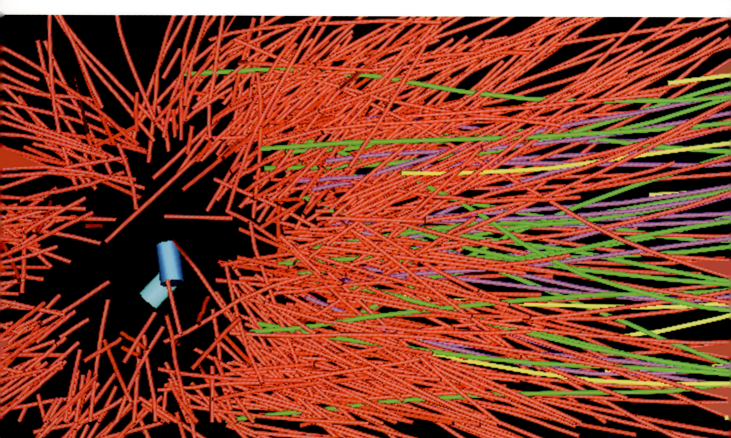

Die Datenflut bewältigen Täglich erhalten Forscher riesige Mengen an Daten aus ihren teils automatisierten Experimenten. Bei der Interpretation der Datenflut hilft die Informatik. Die unzähligen Informationen etwa über Stoffwechsel- und Signalwege in der Zelle können nur noch mithilfe von Computermodellierungen zu einem sinnvollen Ganzen zusammengefügt werden. Verbildlichte Netzwerke erleichtern es, die Komplexität zu erfassen. **Zellen werden zu Fabriken** Mit dem Wissen über Struktur und Funktion ihrer Bestandteile könnten Zellen nachgebaut oder einzellige Organismen neu programmiert werden. Kommt künstliches Leben? Sind alle Prozesse in einer Zelle bekannt, kann der Biologe zum Ingenieur werden. Künstlich erzeugte oder umprogrammierte Organismen sollen in Zukunft etwa Rohöl oder Wasserstoff produzieren und so die Energiekrise lösen. **Bio meets Nano** *Mit Methoden der Nanowissenschaften können heute biologische Systeme untersucht und manipuliert werden. Auch die lebende Natur besteht aus Nanosystemen. Von den eleganten Ideen der Evolution lassen sich Ingenieure der Nanotechnologie inspirieren. Mikrochips, die mit lebenden Zellen kommunizieren und die deren Impulse an einen Computer weiterleiten, verbinden Bio mit Nano. Mikroorganismen, deren Verhalten und Stoffwechsel programmiert werden können, sind die Medikamenten- und Chemikalienproduzenten der Zukunft. Die Abgrenzungen zwischen der belebten und unbelebten Natur verlieren an Schärfe.* **Lebendige Nanostrukturen** Lebewesen bestehen aus Nanobausteinen, ihre Organe enthalten Nanomaschinen: die Kalkhülle einer Muschel, das Holz eines Baumes oder die Zellen einer Hand. Wenn wir die kleinsten Bausteine des Lebens mit physikalischen Methoden untersuchen, verstehen wir, wie sie zusammenwirken. Wir können ihre Funktion auf Makroebene erklären und diese vielleicht sogar beeinflussen. **Leben inspiriert Technik** In der lebenden Welt regulieren ›molekulare Maschinen‹ unzählige Prozesse. Sie können Vorlage für innovative technische Systeme sein. Biologische Strukturen aufzuklären bedeutet jahrzehntelange Arbeit – doch sie lohnt sich. Forscher verstehen Lebens- und Krankheitsprozesse immer besser und können künstliche Nanostrukturen mit spezifischen Eigenschaften konstruieren. **Nano meets Bio** Die Grenzen zwischen lebenden und künstlichen Systemen werden unscharf. Forscher ermöglichen Schnittstellen zwischen beiden Welten. Ingenieure nutzen lebende Systeme als Nanofabriken; auf künstlichen Substraten ordnen sich Nervenzellen kontrolliert entlang vorgegebener Strukturen. Werden die Maschinen des Lebens eines Tages die Welt der Technik erobern?

1 Das Zentrosom ist Ausgangspunkt hunderter Mikrotubuli. Diese Mikroröhrchen verbinden während der Zellteilung (Mitose) die Spindelpole mit den Chromosomenhälften. *Hundreds of microtubules grow outwards from the centrosome. These little microtubes connect the spindle poles with the chromatids during cell division (mitosis).* 2 Das 3D-Modell zeigt eine Halbspindel. *The 3D model shows reconstruction of a half spindle (bottom).* 3 Zellen besitzen ein System der Signalübertragung zur Kommunikation mit ihrer Umgebung. Die zahlreichen Signaltransduktionswege sind hochkompliziert und miteinander vernetzt. *Cells communicate with their environment through a signal transducing system. Signaling pathways are highly complicated and interconnected.* 4 Mit bloßem Auge kann der Betrachter 3D-Darstellungen von Molekülen auf dem Display sehen und mit der Hand bewegen. *Visitors can view three-dimensional images of molecules with the naked eye and turn them with hand movements.*

1, 2 Dr. Thomas Müller-Reichert, Max-Planck-Institut für Molekulare Zellbiologie und Genetik, Dresden 3 ArchiMeDes Movingscience 4 Leihgeber *On loan by* Fraunhofer-Institut für Nachrichtentechnik, Heinrich-Hertz-Institut, Berlin

3

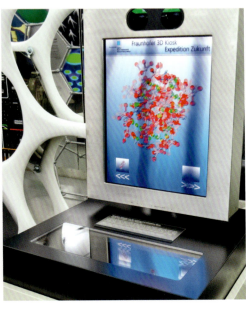

4

The convergence of nanoscience and bioscience

Nowadays, many physical, chemical and biological processes are understood, right down to atomic level. It is becoming possible to manipulate dead and living material on this tiny scale. The boundaries between bio and nano are getting blurred. Researchers are developing atomically precise production processes. Using increasingly sophisticated and complex methods, they decode the syntax and semantics of life – the language of the genes. Integrated circuits are starting to communicate with living cells and to send data to computers. Microorganisms with programmable behaviour and metabolism are emerging: it is undeniable that the convergence of nanoscience and bioscience is forging ahead. Synthetic biology is the new nanotechnology.

›There's plenty of room at the bottom‹ As far back as 50 years ago, the physicist Richard P. Feynmann envisaged that we would ›see‹ and manipulate individual atoms. Over the last thirty years, his dream has come true, for the rules applying to the nano world are different from anywhere else. The nanosciences make the extreme miniaturisation of technology possible. Complicated systems are becoming smaller and smaller – and there's still room to go. However, the tiny components of the nanoworld are still much larger than single atoms or molecules. We have still not arrived at the smallest of all worlds. When will researchers be able to fit single atoms and molecules together perfectly and as required?

Tools with atomic precision The methods used to manipulate material on the nanoscale have developed rapidly. They have led to the birth of the nanosciences. In addition to scanning surfaces and measuring atomic forces, researchers are already able to move individual atoms using an atomic force microscope. However, there is still a long way to go before the construction of functioning nanomachines becomes a reality. **Shakers and movers in nanosystems** There are also tools in the nanoworld: motors, rotors, cutters, joints. However, they have little in common with their equivalents in the macroworld. Up to now, the tiny nano devices have been used to measure extremely small distances or extremely weak forces. In a few years, we might find them in sensors, solar cells, electronic paper or flexible screens. **Art in the nanoworld** We especially need imagination in science. It is not all mathematics, nor all logic, but is somewhat beauty and poetry. *Maria Mitchell (1818–1889), American Astronomer* The examination of nanostructures reveals fantastic images that transport us into fascinating worlds that are full of secrets. Images of the nanocosmos range from mystical grottoes to magical woods. In a mysterious way, the motifs are similar to macroworld objects. Women in nanoscience prove here that they have a creative eye. **What will follow the decoding of the genome?** Entire genomes in higher organisms have been decoded. The molecular description of living systems is becoming increasingly comprehensive. Individual processes can be explained in detail. Nevertheless, we do not know how life works. Rapid technological development is permitting the whole molecular inventory of cells to be catalogued. A tremendous flood of data is presenting biologists with problems that can only be solved with the help of computer specialists and mathematicians. Systems biology is expected to advance from the syntax to the semantics of life. How do millions of molecules work together to create life? Researchers want to predict the behaviour of living systems, and create microorganisms with new properties. **Looking deep inside a cell** High-resolution methods allow us to better understand the interior of a cell. This helps us to a better understanding of the way molecular building blocks function and interact. The structure of our genetic material, our DNA, is also no longer a secret. The genome has been deciphered and is relinquishing its information bit by bit. In the future, everyone will know their own DNA sequence and its significance for their life. **Managing the flood of data** Every day, researchers are presented with huge quantities of data from their sometimes automated experiments. Computer technology helps them to interpret it. The countless pieces

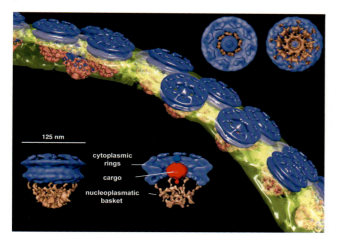
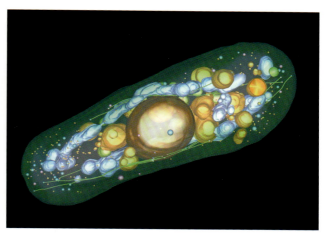

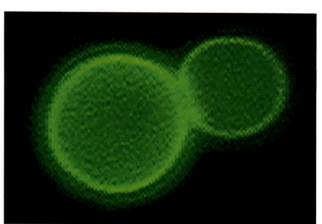
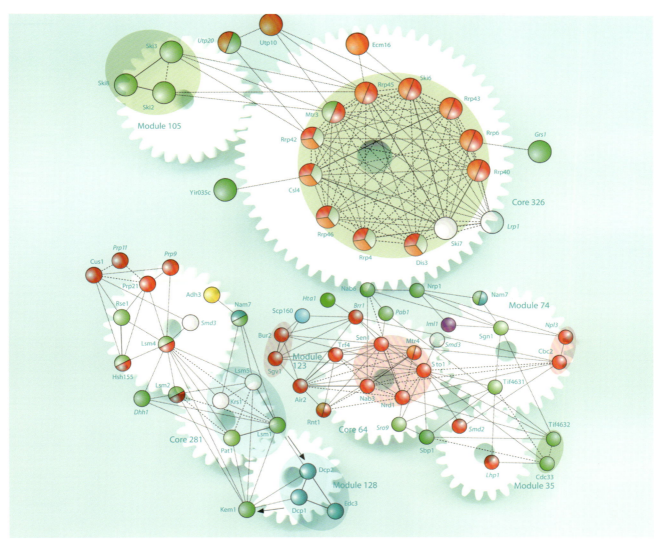

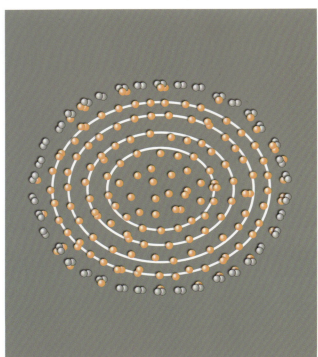
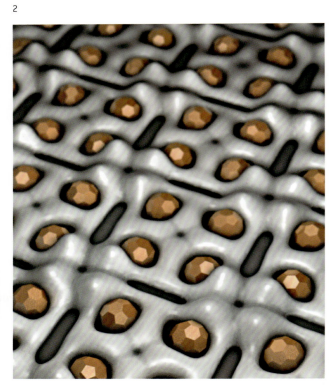
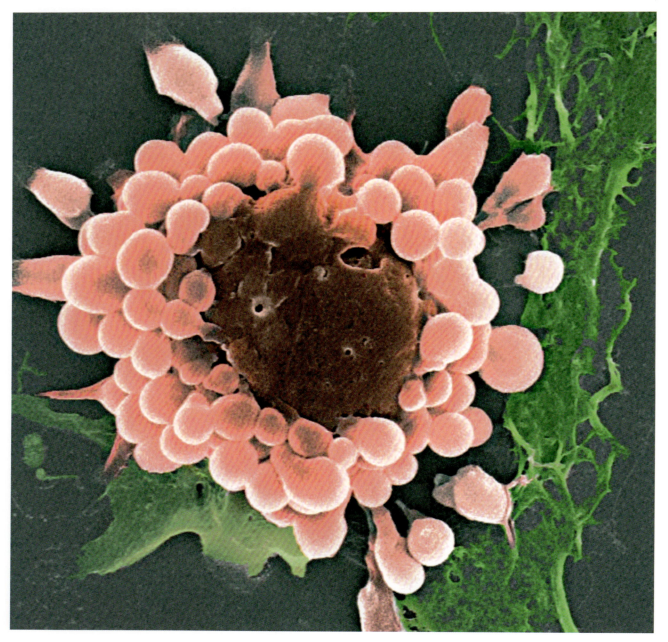

of information about metabolic and signal paths in the cell can only be combined to form a meaningful whole with the aid of computer modelling. Illustrations of networks make it easier to grasp the complexity. **Cells turn into factories** Will knowledge of the structure and the function of their components make it possible to copy cells or to reprogramme single-celled organisms? Is artificial life on the way? Once they know about all the processes in a cell, biologists can become engineers. Artificially created or reprogrammed organisms might, in future, produce crude oil or hydrogen, thus alleviating the energy crisis. **Bio meets nano** *Biological systems can be examined and manipulated via nanoscientific methods. Living nature also consists of nanosystems. Nanotechnology engineers find inspiration in the elegant ideas of evolution. Microchips, which communicate with living cells and conduct their impulses to a computer, link up bio and nano. Microorganisms, whose behaviour and metabolism can be programmed, are the producers of medication and chemicals of the future. The boundaries between animate and inanimate nature are becoming blurred.* **Living nanostructures** Living organisms are made of nanoelements, their organs are nanomachines: the calcium shell of a sea creature, the wood of a tree or the cells of a hand. When we use physical methods to look at the tiniest elements of life, we understand how they interact. We can explain their function at macrolevel – and perhaps even influence it. **Life inspires technology** In the living world, ›molecular machines‹ regulate countless processes. They can be a template for innovative technical systems. It takes decades of work to explain biological structures, but the effort is worthwhile. Researchers are gaining better understanding of life and disease processes and can construct artificial nanostructures with specific characteristics. **Nano meets bio** The frontiers between living and artificial systems are becoming blurred. Researchers create interfaces between the two worlds. Engineers use living systems as nanofactories; on artificial substrates nerve cells arrange themselves along preset structures. Will the engines of the living conquer the world of technology one day?

0,05 mm

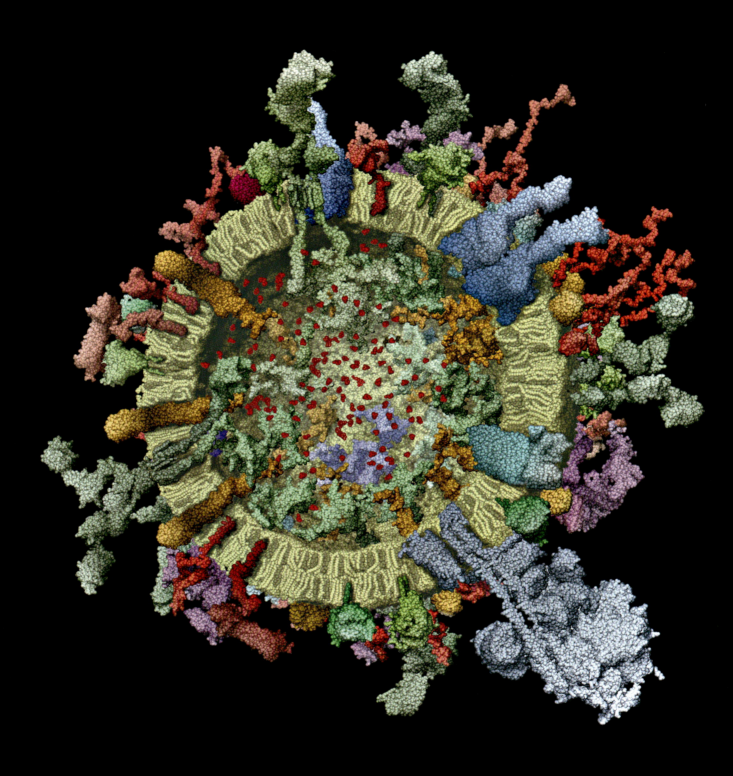

1 Ansicht eines in der Mitte aufgeschnittenen synaptischen Vesikels. Es ist mit einer geringen Konzentration des Neurotransmitters Glutamat (rote Punkte) gefüllt. *A cross-section of a synaptic vesicle. It is filled with a low concentration of the neurotransmitter glutamate (red dots).* —— 2 Nanocytes® – Als winzige Transporter können speziell behandelte (molekular geprägte) Nanopartikel Eiweißmoleküle, hier Insulin, binden und gezielt abgeben. *Nanocytes® – Tiny transporters made from specifically treated (molecular imprinted) nanoparticles bind and deliver proteins such as insulin.* —— 3 Roboter befüllen Mikrotiterplatten in hoher Geschwindigkeit mit Testsubstanzen. Hier können die Besucher mit einer Pipette selbst ausprobieren, wie schnell sie im Vergleich zum Roboter sind. *Robots dispense liquid samples to microtitre plates very rapidly. Here the visitors can test how fast they can pipette compared to a robot.*

1 R. Jahn, Neurobiologie, Max-Planck-Institut für biophysikalische Chemie, Göttingen 2 Leihgeber *On loan by* Fraunhofer-Institut für Grenzflächen- und Bioverfahrenstechnik IGB und Institut für Grenzflächenverfahrenstechnik IGVT, Universität Stuttgart. Wir danken Michael Gantenbrinker für den Export der PDB-Daten von Insulin als STL-Modell. *We would like to thank Michael Gantenbrinker for exporting PDB insulin data as STL model.* 3 Leihgeber *On loan by* Bayer AG

DAS GEHIRN — EIN INTELLIGENTER COMPUTER?
THE BRAIN — AN INTELLIGENT COMPUTER?

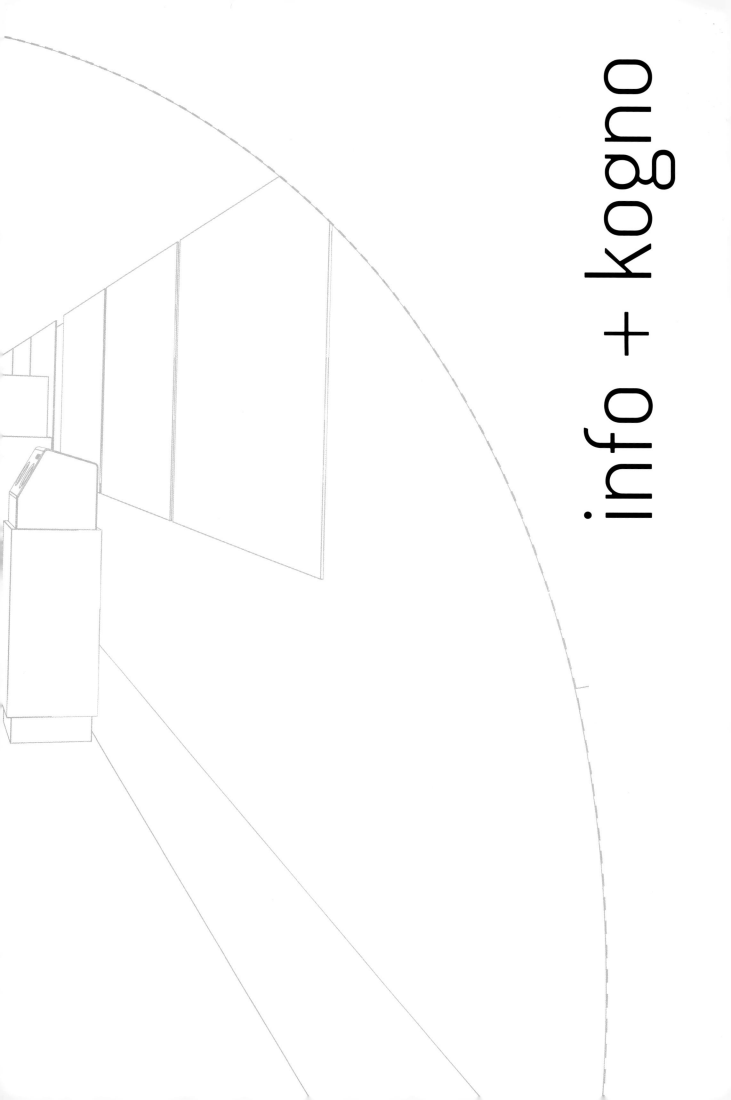

info + kogno

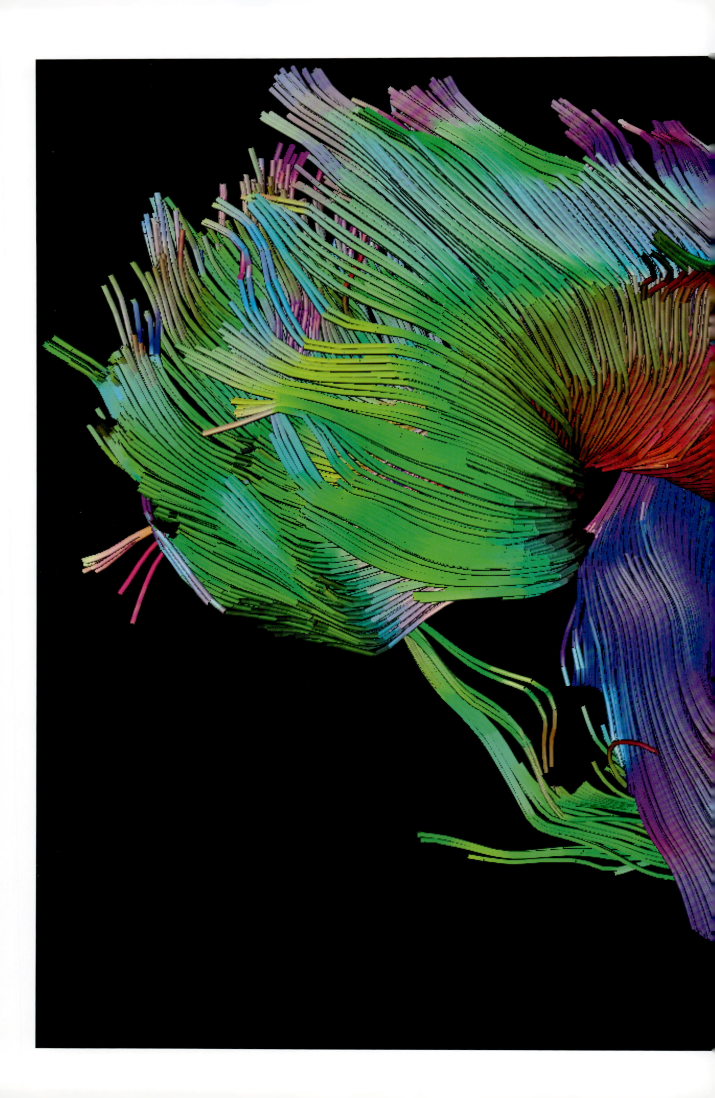

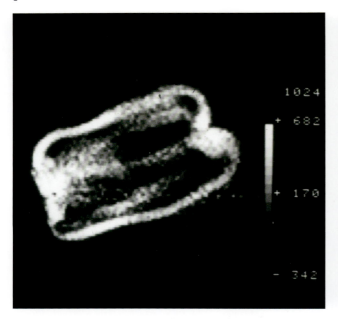
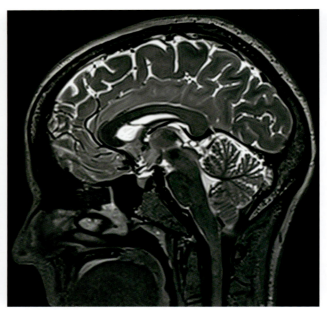
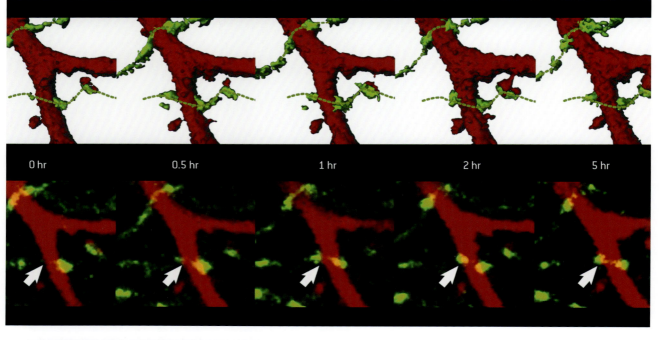

| 0 hr | 0.5 hr | 1 hr | 2 hr | 5 hr |

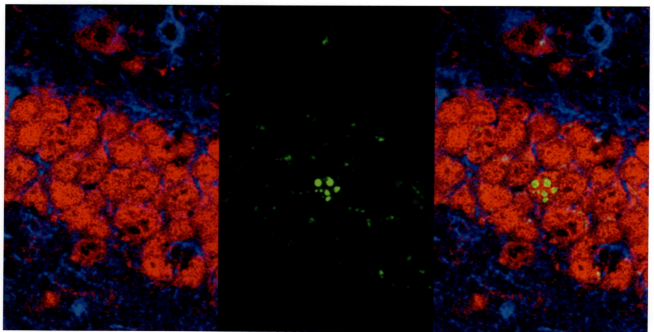

Das Gehirn – ein intelligenter Computer?

Die Funktionen unseres Gehirns werden immer tiefgreifender verstanden. Die Computertechnik entwickelt sich rasant weiter. Wird es je gelingen, die Vorgänge in den Nervenzellen in ihrer Gesamtheit nachzubilden und ein künstliches Gehirn zu schaffen? Mit bildgebenden Verfahren entschlüsseln Forscher die Funktionsweise des Gehirns detailliert auch in Echtzeit. Die Messung der Aktivität ganzer neuronaler Netzwerke verrät uns, wie und was der Mensch denkt und fühlt. Gehirn und Computer können bereits miteinander sprechen. Die zukünftige Computertechnik hilft Systeme zu entwickeln, die wie wir Menschen Informationen verarbeiten und autonom handeln können: Robotern wird Leben eingehaucht. ▬▬ **Der Blick ins Gehirn** *Vielfältige Methoden ermöglichen heute Einblicke in die Arbeitsweise des Gehirns. Wir beobachten die Aktivität ganzer Hirnregionen bis zur Funktion einzelner Neurorezeptoren oder Ionenkanäle. Neuronale Verbindungen behalten ein Leben lang die Fähigkeit zur Anpassung. Die Organisation des Gehirns in Funktionsbereiche ist bekannt. Die Arbeitsweise einzelner Neuronen wird immer besser verstanden. Bildgebende Verfahren stellen den Energiebedarf von Hirnregionen dar und erlauben Rückschlüsse auf einfache geistige Inhalte. Nervenverbindungen werden mikroskopisch beobachtet und neurokognitive Prozesse sogar im Computer simuliert. Dennoch ist weitgehend unbekannt, wie geistige Leistungen durch neuronale Mechanismen erklärt werden können.* ▬▬ **Das Gehirn wird kartografiert** Eine Magnetresonanztomografie (MRT) ermöglicht nicht nur einen Einblick in die Anatomie des Gehirns, sondern auch in dessen Funktionsweise. Dank fortschreitender technischer Entwicklungen werden immer bessere Geräte eingesetzt. Die funktionelle Tomografie zeigt heute sogar, wo im Hirn was passiert. Damit lassen sich Sinneseindrücke und deren Verarbeitung den entsprechenden Gehirnarealen zuordnen. ▬▬ **Wie wir die Welt interpretieren** Bei der menschlichen Wahrnehmung, beispielsweise von Eigenbewegungen oder von der Umwelt, spielen alle Sinne zusammen. Um die komplizierten Mechanismen zu erfassen, werden aufwändige Experimente durchgeführt, die von einer wilden Fahrt auf dem Roboterarm bis hin zur virtuellen Realität reichen. Trotz aller Genialität: Auch unser cleverstes Organ ist vor Täuschungen nicht gefeit. ▬▬ **Wir lernen ein Leben**

lang Was Hänschen nicht lernt, lernt Hans nimmermehr. Diese Redewendung ist widerlegt. Forscher haben bewiesen, dass wir bis ins hohe Alter dazulernen können. Unser Gehirn ist anpassungsfähig. Nervenzellen und Synapsen, die Schaltstellen zwischen diesen Zellen, können neu entstehen. Diese Prozesse kann man selbst, beispielsweise durch ausreichende Bewegung, positiv beeinflussen.

Geist und Gehirn: Die letzten Geheimnisse *Es ist heute vorstellbar, dass grundsätzliche Fragen, wie unser Gehirn Informationen verarbeitet, beantwortet werden. Können die biochemischen und physikalischen Prozesse im Gehirn auch Geist und Bewusstsein erklären? Noch gibt es keine Theorie des Gehirns. Zuerst muss man dynamische Prozesse in Echtzeit erfassen, die in winzigen, hochkomplexen Neuronenschaltkreisen ablaufen. Kombiniert mit Methoden der Kognitionswissenschaften lassen sich diese Ergebnisse mit Hochleistungsrechnern detailliert modellieren. Ist der freie Wille, das Mitgefühl, die moralische Verantwortung, die Entscheidungsfindung oder das Verliebtsein vollständig auf der Grundlage neuronaler Vorgänge zu erklären?* **Struktur und Funktion verstehen** Als Steuerzentrale des Körpers ist das Gehirn unglaublich komplex; ebenso gewaltig ist die Aufgabe, das Zusammenspiel des Zellnetzwerks zu erfassen. Forscher versuchen, mit Computermodellen Vorgänge im Gehirn zu simulieren. Ziel ist es, letztlich jene Mechanismen zu entdecken, die neurologische Krankheiten verursachen. Dabei rücken auch bislang unbeachtete Strukturen in ein neues Licht. **Können wir den Geist erklären?** Das Gehirn wird gescannt, analysiert und am Rechner simuliert. Doch wie kommen etwa Gefühle zustande? Wie lassen sie sich messen oder visualisieren? Buchstaben, Druckerschwärze, Zeichen: Das Material, aus dem fiktive Geschichten gemacht sind, können im Leser reale Gefühle auslösen. Die Kognitive Poetik untersucht die Interaktion zwischen Literatur und ihren Lesern, etwa deren verschiedene emotionale Reaktionen. **Kann man bald Gedanken lesen?** Mithilfe der funktionellen Magnetresonanztomografie ist es in Versuchen bereits möglich, Absichten zu erkennen, bevor diese ausgeführt werden. Aktivitätsmuster im Gehirn können von Computern konkreten Gedanken zugeordnet werden. In Zukunft werden damit Neurotechnologien möglich, die vom Gedankenlesen über Lügendetektion bis hin zur Marktforschung reichen.

Leib und Seele kognitiver Systeme *Das Gehirn ist ein natürliches informationsverarbeitendes System von hoher Komplexität. Wie kann man Wahrnehmung, Lernen, logisches Denken, Entscheidungsfindung, Kommunikation und Handeln simulieren? Die hochgradig vernetzten Nervenzellen verarbeiten die Bits und Bytes im Gehirn. Ihre Wechselwirkungen lassen sich im Computer nachbilden und man kann sogar das Gehirn selbst als Computer betrachten. Dabei gilt immer mehr: Intelligenz benötigt einen Körper. Bereits die Wahl des Körpers, von der Natur meist elegant gelöst, schafft intelligente Eigenschaften, etwa durch mechanische Stabilität. Körper und Geist sind nicht voneinander zu trennen.* **Die Hirn-Computer-Schnittstelle** Werden Elektroden direkt auf das Gehirn gelegt, so kann die Aktivität von Nervenzellen erfasst und an einen Computer übermittelt werden. Hilfe für Schwerstgelähmte: Der Computer übersetzt die Signale der Nervenzellen in der Hirnrinde in die beabsichtigten Bewegungen. Diese könnten an einen PC, Roboterarm, Rollstuhl oder sogar die Muskeln des Patienten weitergegeben werden. **Wie intelligent können Roboter werden?** Schon heute können autonome Roboter mehr, als einfach nur einprogrammierte Aktionen ausführen. Sie agieren, lernen und erfüllen Aufgaben im Team. Roboter verarbeiten und reagieren auf Informationen aus der Umwelt allein auf der Basis von eigenständigen Erfahrungen. Autonome Systeme können sich selbstständig vernetzen und so ganz neue Aufgaben lösen. Wann gelingt der Sprung zur künstlichen Intelligenz? **Die Roboter kommen** Roboter sind aus der Industrie und vielen anderen Bereichen längst nicht mehr wegzudenken. Meist erfüllen sie wenige, wohldefinierte Aufgaben. In vielen Bereichen liefert die Natur

1 Das Gehirn nimmt durch unsere Erfahrungen an, dass sich Gesichter immer nach außen wölben. So sehen wir bei der Hohlmaskenillusion, selbst wenn die Innenseite der Maske betrachtet wird, immer ein uns zugewandtes Gesicht. *The brain knows from experience that faces always curve outward. That's why the face in the hollow mask illusion is always facing towards you even when you're looking at the inside of the mask.* 2 Bildgebende Verfahren wie die funktionelle Magnetresonanztomografie können zeigen, welche Gehirnregionen bei Bewegungen oder Aktionen, wie beispielsweise beim Sprechen, aktiv sind. *Movements or actions like speaking cause activity in certain regions of the human brain. These can be visualised by medical imaging techniques such as functional Magnetic Resonance Imaging.* 3 Gliazellen sorgen für die elektrische Isolierung und Versorgung der Nervenzellen. Sie spielen aber auch eine wichtige Rolle in der Verarbeitung und Weiterleitung von Informationen und in der Reaktion des Gehirns auf Verletzungen. *Glial cells provide electrical insulation and metabolic and trophic support for neurons. But they also play an important role in processing information and in the reaction of the brain to injuries.* 4 Dieser Netzwerkchip besteht aus 384 künstlichen Nervenzellen und 100.000 Synapsen. Damit könnten Computer nach Prinzipien des menschlichen Gehirns vielleicht eines Tages möglich sein. *This network chip consists of 384 artificial neurons and 100,000 synapses. Computers patterned on the human mind might one day be possible.* 5 In einer Studie nahm die Aktivierung in bestimmten Teilen des seitlichen Frontalhirns (Pfeil) zu, wenn der Proband ein wiedererkanntes Detail verheimlichte. *In a study, activity in certain parts of the lateral frontal lobe (arrow) increased when the test person sought to conceal a detail he had recognised.* 6 Durch die Verheimlichung stieg die Hautleitfähigkeit (Pfeil) und könnte somit an der Schweißdrüsenregulation beteiligt sein. *Through this concealment, the conductivity of the skin rose sharply. This caused activation of the affected region (arrow), which could thus be involved in the regulation of the sweat gland.*

1 Leihgeber *On loan by* Max-Planck-Institut für biologische Kybernetik, Tübingen 2 Leihgeber *On loan by* Max-Planck-Institut für Kognitions- und Neurowissenschaften, Leipzig 3 Stefanie Robel, Arbeitsgruppe Magdalena Götz, Physiologisches Institut, Ludwig-Maximilians-Universität München und Institut für Stammzellforschung, Helmholtz Zentrum München, Deutsches Forschungszentrum für Gesundheit und Umwelt (GmbH) 4 Universität Heidelberg, FACETS-Projekt 5, 6 Mit freundlicher Genehmigung von Matthias Gamer

1

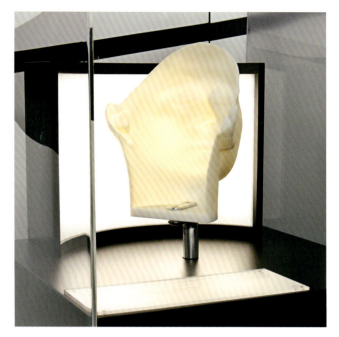

2

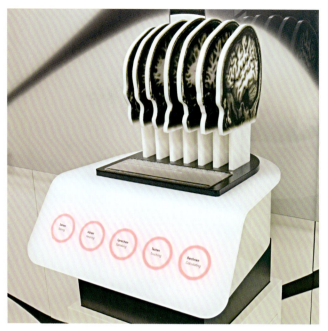

3

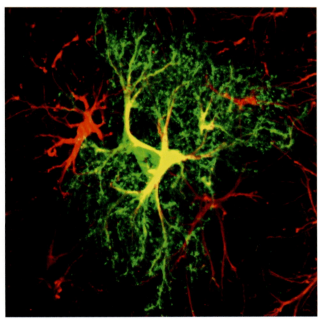

4

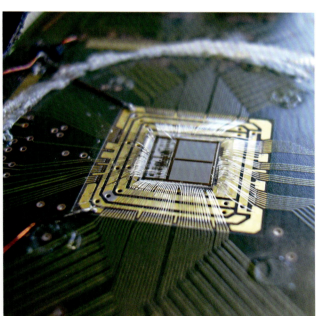

5

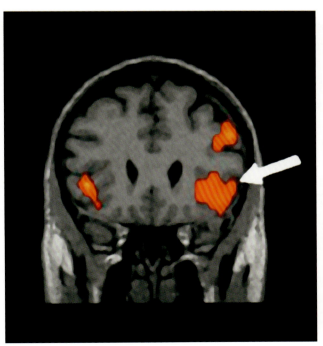

6

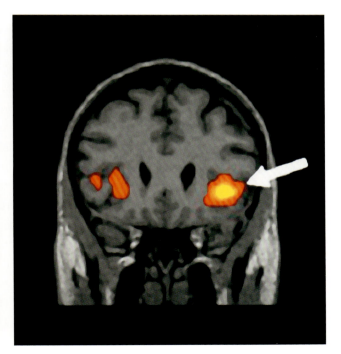

Ideen, etwa für die Fortbewegung. Doch das Zusammenspiel von Mensch und Maschine ist oft noch schwierig: Robotern fällt es schwer, Sprache und Gesten zu verstehen, oder sich in einer komplexen Umgebung zu orientieren. **Wohin entwickelt sich die Computertechnik?** *Die Realisierung künstlicher Intelligenz und kognitiver Systeme wird nur mit fortschrittlichster Computertechnik gelingen. Die Anforderungen an die Leistungsfähigkeit zukünftiger Hard- und Software steigen, doch innovative Technologien halten Schritt. Zukünftige Computer werden noch schneller. Eine größere Integrationsdichte führt zu höheren Taktfrequenzen. Auch bisher nicht genutzte physikalische Effekte lassen sich zur Steigerung der Leistungsfähigkeit einsetzen oder führen zu neuartigen Bauprinzipen: Licht eignet sich zur Signalverarbeitung in einem optischen Computer. Biocomputer nutzen Erbsubstanz als Speicher und Verarbeitungsmedium. Der Quantencomputer rechnet mit überlagerten Zuständen massiv parallel.* **Kleiner, schneller, universeller** Computerchips werden immer kleiner; ihre Miniaturisierung hat bisher jede technische Hürde überwunden. Wie lange kann die Entwicklung noch andauern? Besteht ein Transistor nur noch aus wenigen Atomen, ist eine prinzipielle Grenze erreicht. Einzelflußquantenschaltungen, die quantenmechanische Effekte in supraleitenden Materialien nutzen, versprechen noch schnellere Chips. **Überlagerte Bits und Bytes** In der Quanteninformatik können die ersten Qubits, die kleinsten Speichereinheiten, bereits rechnen – der Einsatz eines Quantencomputers rückt damit näher. In normalen Computern sind Bits entweder 0 oder 1. Qubits werden durch überlagerte Quantenzustände realisiert und können daher die Werte 0 und 1 gleichzeitig annehmen. Dadurch sind viel effizientere, massiv parallele Berechnungen möglich.

The brain – an intelligent computer?

Our understanding of how the human brain functions is deepening all the time. And computer technology is developing in leaps and bounds. Will scientists ever manage to fully replicate what happens in our nerve cells and create an artificial brain? Using imaging technology, scientists are deciphering the processes that make our brain work – in detail and even in real time. Measuring the activity of entire neural networks discloses how and what we think and feel. Brain and computer are already able to speak to one another. The computer technology of the future is helping to develop systems capable of processing information and acting autonomously, as we do. It is breathing life into robots.

Looking into the brain *Nowadays, there are numerous ways for us to get an idea of how the human brain works. We can observe the activity of entire regions of the brain or look at the function of single neuro-receptors or ion channels. Neural networks retain their ability to adapt throughout an entire lifetime. It is known that the brain is organised into functional areas. Our understanding of how individual neurons work is growing all the time. Imaging techniques reveal the energy used by different regions of the brain and allow scientists to infer some simple mental contents. Synapses are examined under the microscope and neuro-cognitive processes are even simulated on the computer. Yet we are still unable to explain the feats of mental agility our brain can perform simply by examining the neural mechanisms involved.*

Mapping the brain Magnetic resonance imaging (MRI) not only provides insight into the anatomy of the brain; it also gives us an idea of how the brain works. Progressing technical developments ensure that the quality of imaging devices is constantly improving. Functional tomography is already able to show us what happens where in the brain. This enables scientists to match sensory input with the areas of the brain where it is processed.

How do we interpret the world *Human perception, for example of own movements or of the environment, is a result of the interplay of all senses. In order to comprehend the complex mechanisms intricate experiments must be performed, reaching from a wild journey on the lever of a robot up to virtual reality. Despite all ingenuity: Even our smartest organ is not immune to delusion.*

Life-long learning You can't teach an old dog new tricks, or so they say. Scientists have disproved this proverb. We now know that people can continue to learn into old age. Our brain is adaptable. Nerve cells and synapses, the switch points between the cells, can be reformed. These are processes that we ourselves can influence constructively, for instance by getting enough exercise.

Mind and brain: The last remaining secrets *It is quite conceivable that we may one day be able to answer the fundamental questions of how our brain processes information. Can the biochemical and physical processes in the brain also explain the human mind and consciousness? There is, as yet, no theory of the brain. First, scientists need to capture dynamic processes in real time, processes that take place in tiny, highly complex neural networks. In combination with methods drawn from the cognitive sciences, supercomputers can model these findings in detail. Will it be possible to explain our free will, sympathy, morals, decisionmaking or the feeling of being in love, purely on the basis of neural processes?*

Understanding structure and function As the body's control centre, the brain is incredibly complex; the task of understanding the interactions within the cell network is equally complicated. Scientists use computer modelling techniques in an attempt to simulate processes in the brain. What they are ultimately trying to do is discover the mechanisms that cause neurological diseases. Structures that have thus far gone unnoticed are now being seen in a new light.

Can we explain the mind? The brain is scanned, analysed and processes are modelled by computers. But how are emotions generated? How can they be measured or visualised? Letters, printing ink, characters: The reading of fiction elicits real feelings. Cognitive Poetics investigates the interplay between literature and its readers, for

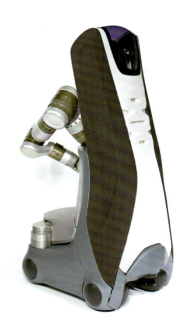

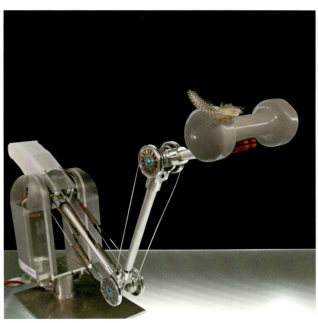

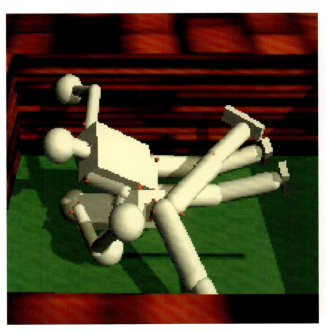
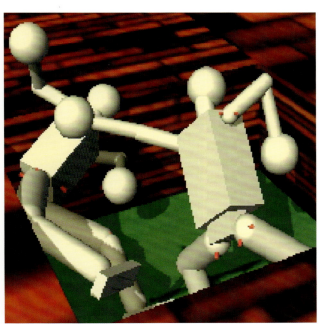

1

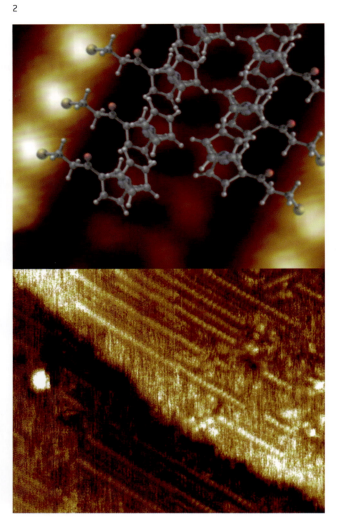

2

3

example their diverse emotional reactions. **Will it soon be possible to read thoughts?** With the use of functional magnetic resonance imaging, it is already possible for scientists in the lab to detect intentions before they are carried out. Computers can allocate patterns of activity in the brain to specific thoughts. In future, this will enable the development of neuro-technologies capable of doing anything from mind reading and lie detection to market research. **Body and soul of cognitive systems** *The brain is a natural information-processing system of extreme complexity. How can we simulate perception, learning, logical thinking, decision making, communication and action? The intensely cross-linked nerve cells process the bits and bytes in the brain. Their interactions can be replicated on the computer and the brain itself can even be seen as a computer. These days, it's truer than ever that intelligence needs a body in which to function. The very choice of body, for which nature usually finds an elegant solution, creates intelligent characteristics, by providing mechanical stability, for instance. Body and mind are inseparable.* **The brain-computer interface** If electrodes are attached directly onto the brain, nerve cell activity can be recorded and transmitted to a computer. Help for the severely paralysed: The computer translates the signals from nerve cells in the cerebral cortex into the intended movements. These could be transmitted to a computer, a robotic arm, a wheelchair or even the muscles of a patient. **Just how intelligent can robots become?** Autonomous robots can already do more than merely execute programmed actions. They can act, learn and fulfil tasks in a team. Robots process and react to information from the environment exclusively on the basis of their own experiences. Autonomous systems can form networks independently and thus solve brand new tasks. When will we make the leap to artificial intelligence? **The robots are coming** Nowadays, it's impossible to imagine industry and many other sectors without robots. They usually carry out a small number of well-defined tasks. It is nature that provides the ideas in many areas, such as locomotion. But the interaction between man and machine is often difficult: Robots have problems understanding language and gestures or finding their bearings in a complex environment. **Where is computer technology heading?** *The realisation of artificial intelligence and cognitive systems will only succeed with the most advanced of computer technologies. The demands on the performance of future hardware and software are increasing – however, innovative technologies are keeping up. The computers of the future will be even faster. Greater integration density brings higher tact frequencies. As yet unexploited physical effects can also be used to increase performance or facilitate new constructional principles: Light lends itself to the processing of signals in an optical computer. Biocomputers use genetic substance as a memory and processing medium. Quantum computers relying on superimposed quantum states form massively parallel computers.*

Smaller, faster, more universal Computer chips are becoming ever smaller; the process of miniaturisation has so far overcome every technical hurdle it has faced. How much longer can this development go on? Once a transistor consists of no more than a few atoms, a theoretical limit has been reached. Single-flux quantum circuits, which exploit the quantum-mechanical effects of superconducting materials, hold the promise of even faster chips. **Superimposed bits and bytes** In quantum informatics, the first qubits, the smallest units of information, are already capable of computing – bringing quantum computers that much closer. In normal computers, bits are either 0 or 1. Qubits are realised by superimposed quantum states and can therefore accept the values 0 and 1 simultaneously. This enables much more efficient, massively parallel calculations to be carried out.

1 Zusammen mit Team NimbRo gewann ›Paul‹ das Finale der ›Humanoiden‹-Liga beim RoboCup 2007. Bei diesem Wettbewerb werden Aspekte der Intelligenz, Köperbeherrschung und Teamspiel untersucht. *Together with the NimbRo Team, ›Paul‹ won the final of the ›Humanoid‹ League in the 2007 RoboCup. The contest examines aspects of intelligence, body control and team play.* 2 Diese elektroaktiven Moleküle ordnen sich selbstständig zu Reihen. Molekulare Elektronik könnte in zukünftigen elektronischen Geräten den Energieverbrauch minimieren. *These electro-active molecules arrange themselves in rows autonomously. Molecular electronics could minimise energy consumption in future electronic devices.* 3 Das europäische SECOQC-Projekt nutzt die Quantenkryptografie für sichere Kommunikation. Der Netzwerk-Prototyp aus sechs Knoten wurde in Wien getestet. *The European SECOQC project uses quantum cryptography for secure communication. The six-node prototype network was tested in Vienna.*

1 Leihgeber *On loan by* NimbRo, AG Autonome Intelligente Systeme, Universität Bonn 2 JARA – Forschungszentrum Jülich & RWTH Aachen 3 Austrian Research Centers

AUF DEM WEG IN EINE DIGITALE GESELLSCHAFT
ON THE ROAD TO A DIGITAL SOCIETY

vernetzt + global

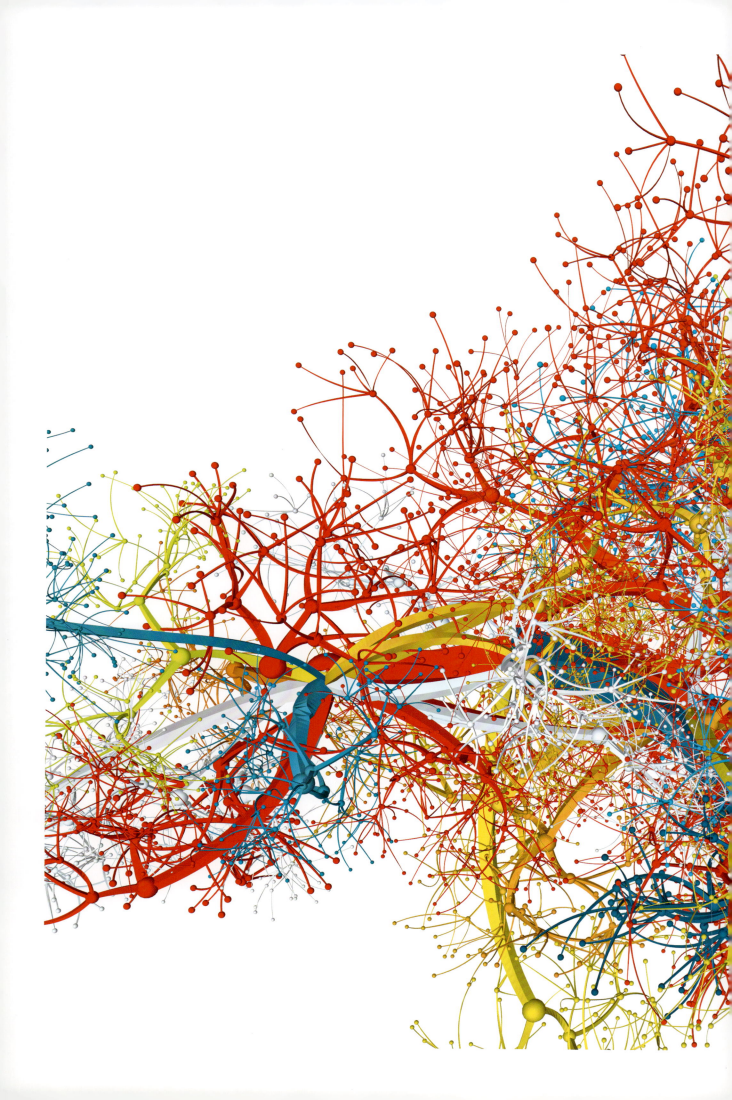

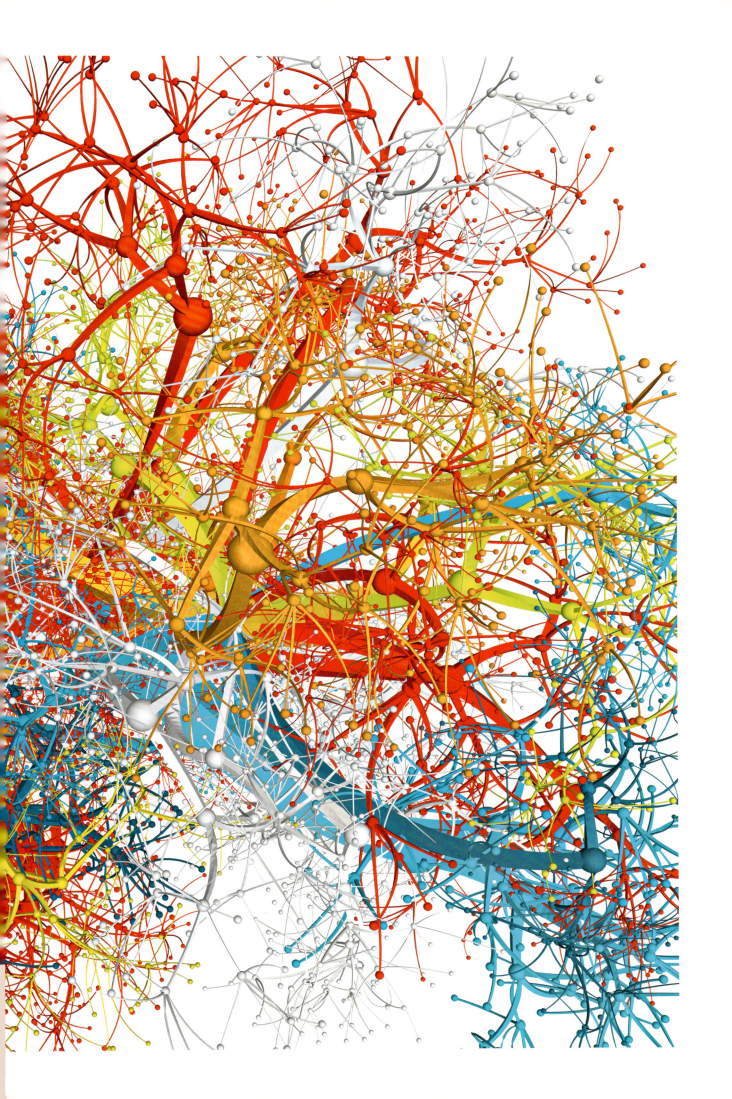

Die Entwicklung des Internets *Internet Evolution*

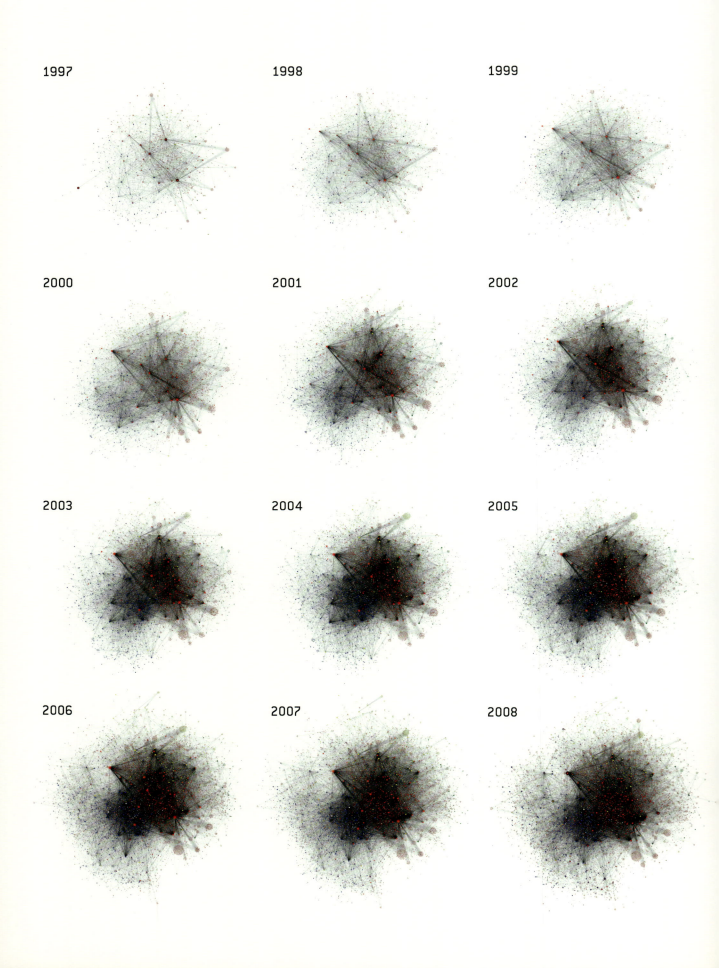

Auf dem Weg in eine digitale Gesellschaft

Computer durchdringen alle Bereiche unserer Welt, schnelle Netze umspannen den Globus. Das verändert die Art, wie wir leben, arbeiten und kommunizieren. Eine digitale Welt entsteht, in der sich Kreativität und Innovation auf neue Weise entfalten können. In der Folge beeinflussen Wissenschaft und Forschung unser Leben im 21. Jahrhundert stärker als jemals zuvor. Dafür sorgen massive Investitionen in Forschung und Entwicklung, aber auch intensive Kooperation und harter Wettbewerb. Das Zusammenwachsen von Nano-, Bio-, Informations- und Neurotechnologien ermöglicht völlig neue Anwendungen. Wissen wird – neben Boden, Kapital und Arbeit – zum entscheidenden Faktor für Wohlstand, aber auch für die Lösung globaler Probleme. Dabei müssen digitale Freiheit und digitale Sicherheit harmonieren. ____ **Wissenschaft 2020: Die systematische Vermessung der Welt** *Millionen Wissenschaftler ergründen unsere Welt heute im Detail, quer über alle Skalen von Raum, Zeit, Energie oder Komplexität. Grundlegend neue Erkenntnisse kommen etwa aus der Erforschung von Themen zwischen den Disziplinen oder extremen Materiezuständen. Wissenschaft verharrt längst nicht mehr im Bereich unserer natürlichen Lebensumstände und Wahrnehmungen. Beträchtliche Mittel werden investiert, um die kleinsten Bausteine unserer Welt zu entschlüsseln und zu verstehen, wie ihr Zusammenwirken ganz neue Eigenschaften hervorbringt. Innovationstreiber der Forschung sind heute: Datenerfassung über digitale Sensoren, Speicherung, Auswertung und Visualisierung über Computer und Software sowie weltweiter Informations- und Wissensaustausch.* ____ **Der Aufwand für neues Wissen steigt** Heute gibt es keinen Bereich unseres Lebens, der nicht erforscht wird. Zugleich wird es immer aufwändiger, neues Wissen zu produzieren. Neue Forschungsmethoden und -technologien ermöglichen es uns heute, auch die ›äußersten Grenzen‹ der Welt zu erforschen: extrem schnelle oder langsame Prozesse, allerkleinste Bausteine und allergrößte Strukturen, extrem kalte oder heiße Zustände. ____ **Vernetztes Wissen stellt sich globalen Herausforderungen** *Dank weltumspannender Informations- und Kommunikationsnetze werden uns die Herausforderungen, vor denen unsere Zivilisation langfristig steht, schneller und deutlicher bewusst. Wir können früher damit beginnen, gemeinsame Lösungen zu entwickeln. Forschung erfolgt bei vielen Themen global – in enger Kooperation oder im weltweiten Wettbewerb um die schnellste und beste Lösung. Nationale Grenzen verlieren an Bedeutung. Millionen von Wissenschaftlern arbeiten über Länder, Kontinente und Zeitzonen hinweg in Tausenden von Laboratorien. Ihre weltweite Vernetzung erhöht die Vielfalt und Effizienz von Wissenschaft und Technik. Das wiederum verstärkt die Globalisierung und Vernetzung. In einer sich derart verändernden Welt muss jedes Land seinen Platz neu finden.* ____ **Das Ende der Distanz** Die Menschheit steht lokal wie weltweit vor gewaltigen Herausforderungen – Ressourcen nachhaltig zu nutzen und eine globale Ökonomie zu organisieren. Komplexe Prozesse werden rund um den Globus im Detail erfasst, in Datenbanken gesammelt und in Computer-Netzwerken ausgewertet. Neue Visualisierungstechniken ermöglichen es, immer größere Datensätze zu analysieren und daraus Schlussfolgerungen zu ziehen. ____ **Globale Vernetzung als Triebkraft der Wissenschaft** Die Menschheit steht lokal wie weltweit vor gewaltigen Herausforderungen – Ressourcen nachhaltig zu nutzen und eine globale Ökonomie zu organisieren. Komplexe Prozesse werden rund um den Globus im Detail erfasst, in Datenbanken gesammelt und in Computer-Netzwerken ausgewertet. Neue Visualisierungstechniken ermöglichen es, immer größere Datensätze zu analysieren und daraus Schlussfolgerungen zu ziehen. ____ **Globale Gerechtigkeit heißt Wissen zu teilen** Wissen bedeutet Entwicklung. Die digitalen Technologien ermöglichen es den Menschen heute überall auf der Welt, neues Wissen zu nutzen. Damit sich die Wissenskluft zwischen reichen und armen Ländern verringert und die Chancengleichheit verbessert, benötigen gerade die Menschen in den am wenigsten entwickelten Ländern Zugang zum Internet. Im Web wird mehr und mehr neues Wissen frei zugänglich.——**Entsteht ein globales Gehirn?** *Globale Netzwerke von Wissenschaftlern, schnelle Computer und intelligente Software helfen dabei, komplexe Zusammenhänge besser zu verstehen. Die digitale Vernetzung der Vielen steigert die Kreativität des Einzelnen – wir leben in ›beschleunigten Zeiten‹. Die Investitionen in die Forschung steigen weltweit. Messplätze, Forscher und Wissen werden immer intelligenter miteinander vernetzt, Zusammenhänge früher erkannt. Wir sind Zeugen einer Explosion der menschlichen Kreativität. Treiber dieses Wandels sind digitale Informations- und Kommunikationstechnologien. Neues Wissen entsteht heute auf der ganzen Welt, es wird künftig überall verfügbar sein. Entscheidend ist dann, wer es schnell, kompetent und bedarfsgerecht zu nutzen weiß.*——**Wissen nach Bedarf** Das Internet der Zukunft wird zur ›Antwort-Maschine‹, zum persönlichen Assistenten, Berater, verlängerten Gedächtnis, Weg- und Lebensbegleiter. Unser Verhältnis zum Wissen verändert sich. Das Internet wird zur ›Wissenswerkstatt‹: Immer neue Werkzeuge helfen, vielfältige Wissensquellen zu erschließen, Erfahrungen auszutauschen oder sich über den ›Zeitgeist‹ just in time informieren zu lassen.——**Wie entstehen neue Qualitäten?** Mit leistungsfähigen Computern und Datennetzen gelingt es schrittweise zu verstehen, warum das Ganze oft mehr ist als die Summe seiner Teile. Analyseroboter, stetig steigende Rechenleistung, angewandte Mathematik und schnelle Datennetze – damit gelingt es schrittweise zu verstehen, wieso das Zusammenwirken einfacher Bausteine völlig neue Eigenschaften oder ein ganz neues Verhalten produzieren kann.——**Wissenschaftliche Freiheit – Elixier der Zivilisation** *Der wissenschaftlich-technische Fortschritt hängt davon ab, über welche Freiräume Wissenschaftler verfügen und wie offen sie sich austauschen können. Triebkraft des Fortschritts ist die Kreativität des Einzelnen. Diese wächst heute dank der weltweiten Vernetzung von Wissenschaftlern. Im Ergebnis werden Projekte schneller gestartet, neue Ideen rascher aufgegriffen.*——**Was bringt uns die digitale Zukunft?** *Doch nicht nur Wissen, auch Waren und Dienstleistungen vernetzen sich digital – die Grenzen zwischen unseren Arbeits- und Lebenswelten, zwischen realem und virtuellem Leben verwischen. Forschung verändert unser Leben. In rascher Folge entstehen weltweit neue Forschungsergebnisse, die jedoch nicht automatisch angewandt werden. Innovationen müssen ihren Weg in eine hoch vernetzte, für Veränderungen sensibel gewordene Gesellschaft finden. Entscheidend ist auch, wie offen eine Gesellschaft für Neues ist und von welchen Bedürfnissen und Wertvorstellungen sich die Menschen leiten lassen. Neues Wissen kann brachliegen oder auch schneller von anderen aufgegriffen werden.*——**Wie werden wir leben?** Ein neues, weltweites Informations- und Kommunikationssystem verändert unser Denken und unsere Kultur grundlegend. Schrittweise vernetzen sich Dinge und Dienste über winzige Einheiten miteinander. Alles wird ›smart‹, ›intelligent‹ oder ›virtuell‹, sei es im Heimbereich, in Produktion, Mobilität und Logistik oder in der Gesundheitsversorgung und Energiewirtschaft.——**Wohin entwickelt sich das Internet?** Beim Internet der Zukunft geht es um mehr als Zugang, Bandbreite und Endgeräte. Sein Potenzial liegt in neuartigen Anwendungen und digitalen Inhalten. Das Internet der Zukunft verbindet Informationen, Waren, Dienstleistungen und Menschen miteinander. Das semantische Web hilft, das Richtige zu finden. Vernetzte Systeme werden uns auf Schritt und Tritt begleiten. Dinge und Dienste um uns herum sprechen miteinander und mit uns.

1 Die Visualisierung zeigt die Bewegung von Flugzeugen im amerikanischen Luftraum. Die Daten stammen von der Federal Aviation Administration. *The visualisation shows the movement of planes in American airspace. The data was provided by the Federal Aviation Administration.*——2 Soziale Netzwerke im Internet werden immer wichtiger. Autoren arbeiten in so genannten Collaboration Networks zusammen, die ständig wachsen. *The social web space gains ever more importance. Authors work together in so called collaboration networks, which continue to grow.*——3 Querverweise in der Bibel. Das Balkendiagramm im unteren Teil der Visualisierung zeigt die einzelnen Kapitel der Bibel, Bücher wechseln sich in weiß und hellgrau ab. Alle 63.779 Querverweise sind durch einen Bogen dargestellt. *Bible Cross-References. The bar graph along the bottom shows all chapters of the Bible. Books alternate in colour between white and light grey. All 63,779 cross-references are depicted by single arcs.*

1 Aaron Koblin 2 Universität Stuttgart, SONIVIS, www.sonivis.org 3 Chris Harrison

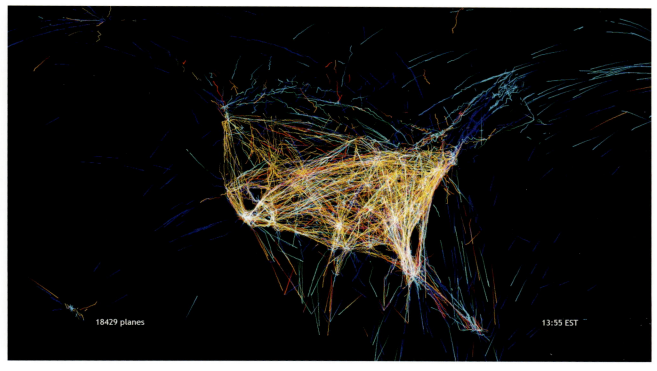

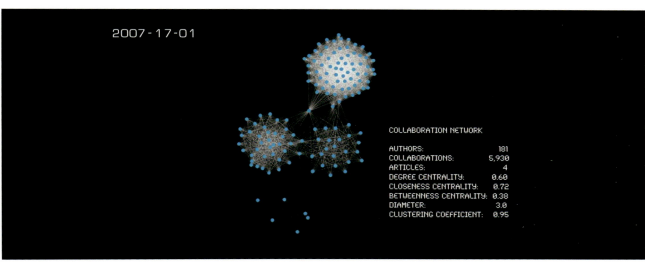

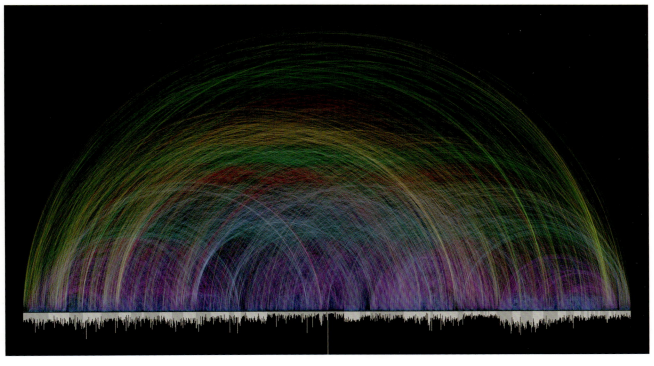

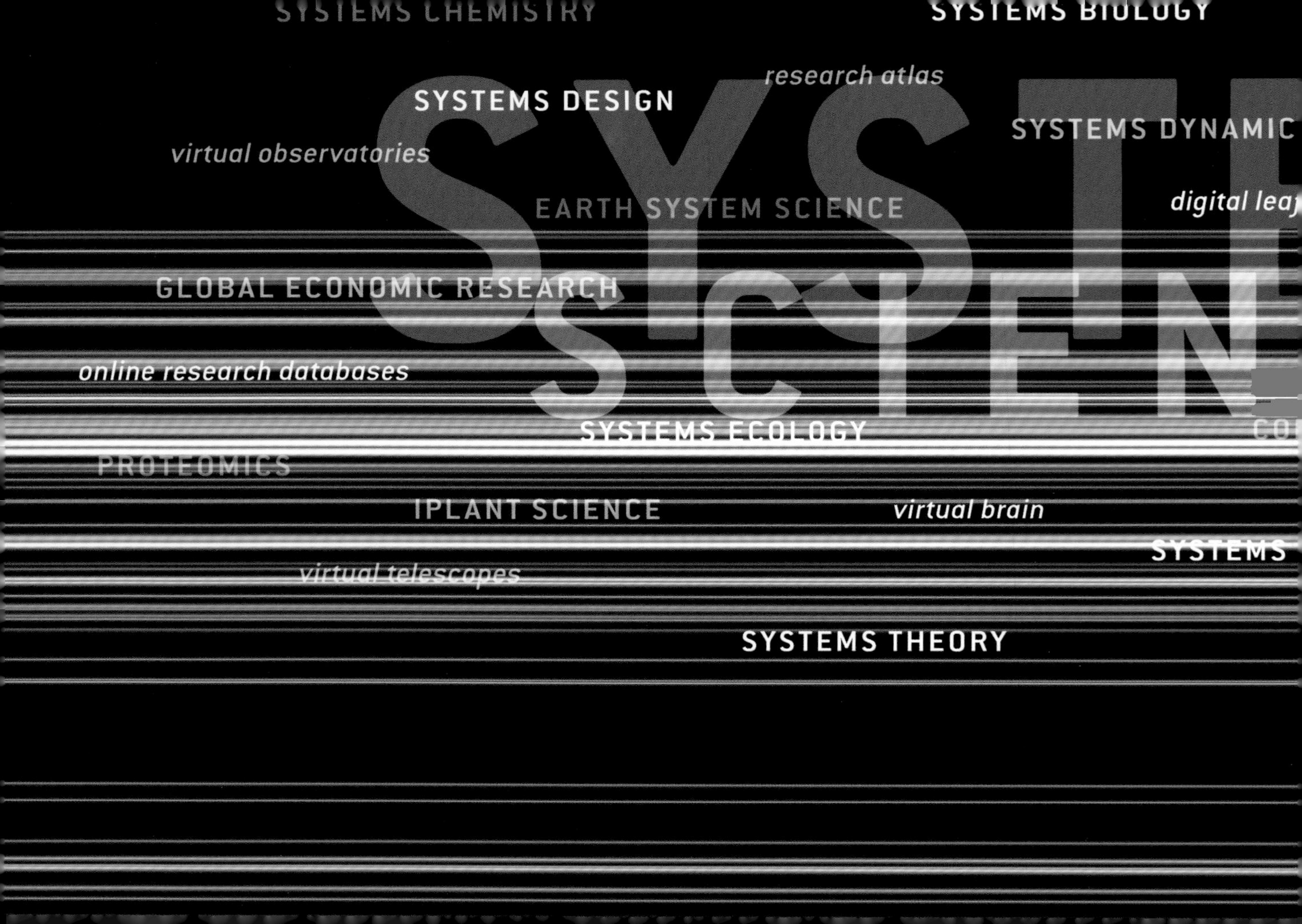
SYSTEMS CHEMISTRY
SYSTEMS BIOLOGY
research atlas
SYSTEMS DESIGN
SYSTEMS DYNAMIC
virtual observatories
SYSTE
EARTH SYSTEM SCIENCE
digital lea
GLOBAL ECONOMIC RESEARCH
SCIEN
online research databases
SYSTEMS ECOLOGY
PROTEOMICS
IPLANT SCIENCE
virtual brain
SYSTEMS
virtual telescopes
SYSTEMS THEORY

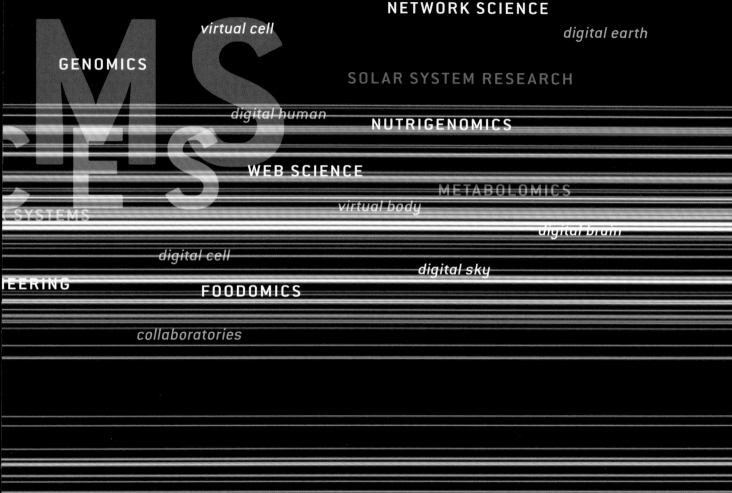

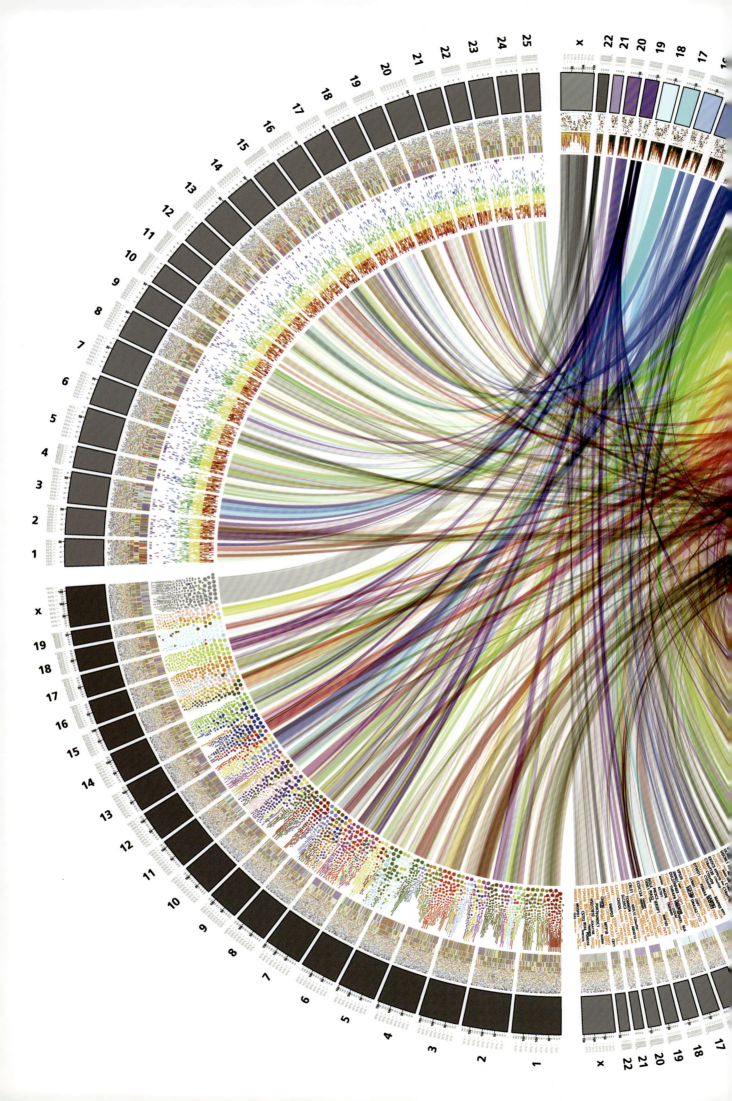

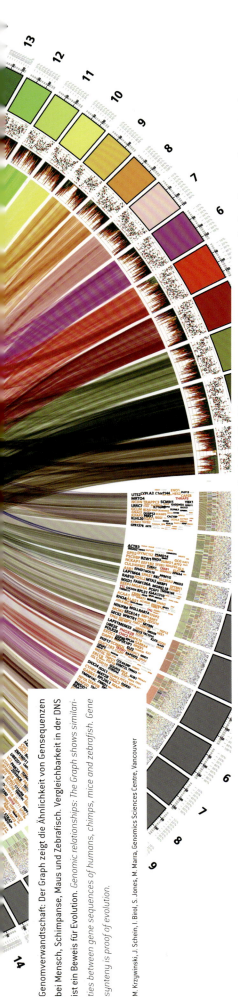

Genomverwandtschaft: Der Graph zeigt die Ähnlichkeit von Gensequenzen bei Mensch, Schimpanse, Maus und Zebrafisch. Vergleichbarkeit in der DNS ist ein Beweis für Evolution. *Genomic relationships: The Graph shows similarities between gene sequences of humans, chimps, mice and zebrafish. Gene synteny is proof of evolution.*

M. Krzywinski, J. Schein, I. Birol, S. Jones, M. Marra, Genomics Sciences Centre, Vancouver

On the road to a digital society

Computer technology is an ubiquitous element of our world, and fast networks are spanning the globe. This is changing the way we live and work and communicate. A new digital world is emerging, an environment in which creativity and innovation can flourish in many new ways. As a result, science and research have a greater influence on our life in the 21st century than ever before. This is attributable to massive investments in research and development, but also to intensive cooperation and tough competition. The convergence of nano-, bio-, information- and neurotechnologies facilitates completely new applications. Taking its place beside the more traditional factors of land, capital and employment, knowledge is fast becoming the decisive factor for prosperity – and also for the resolution of global problems. In this, the appropriate balance between digital freedom and digital security must be maintained.

Science 2020: Systematically surveying the world *Millions of scientists are getting to the bottom of the secrets of our world, across the whole spectrum of space, time, energy and complexity. Fundamentally new knowledge is emerging from research into inter-disciplinary topics or extreme states of matter. Science long ago escaped the constraints of working only in the realm of our natural living conditions and our perceptions. Considerable investment is flowing into efforts to decode the smallest building blocks of our world and to understand how their interplay produces brand new qualities. The drivers of innovation in research today are data capture via digital sensors; storage, analysis and visualisation via computer and software; and the global exchange of information and knowledge.*

The cost of new knowledge is rising There is now no part of our life that is not the subject of research. At the same time, it is becoming ever more difficult to generate new knowledge. These days, new research methods and technologies enable us to study even the ›farthest frontiers‹ of the world: extremely fast or slow processes, the tiniest building blocks or the largest structures, extreme cold or extreme heat.

Networked knowledge takes on global challenges *Thanks to worldwide information and communication networks, the challenges our civilisation faces in the long term are known to us sooner and more clearly than ever before. We can start developing solutions together at an earlier stage. Research on many topics is global – taking place in close cooperation or in international competition for the fastest and best solutions. National boundaries are becoming irrelevant. Millions of scientists work across countries, continents and time zones in thousands of labs. Their global networking enhances the diversity and efficiency of science and technology. And this, in turn, reinforces globalisation and networking. In a world changing at such a pace, each country must redefine its place.*

The end of distance Mankind faces enormous challenges both locally and globally – the challenge of using resources sustainably and of organising a global economy. Across the globe, complex processes are being recorded in detail, collated in databases and analysed in computer networks. New visualisation techniques make it possible to analyse larger and larger data records and to draw conclusions from the results.

Global networking as the driving force of science In the early days, the Internet linked up scientists, large-scale equipment and information; now it networks computational power and enormous amounts of data through grid and cloud computing. A global Semantic Web is emerging, bringing together data, expertise and knowledge that had previously been distributed among virtual libraries and observatories. The information is being intelligently developed, new forms of cooperation are arising, and research is becoming more productive and efficient.

Global equality means sharing knowledge Knowledge means development. Digital technologies enable people all over the world to use new knowledge. In order to bridge the knowledge divide between rich and poor nations and improve equal opportunities, it is those in the least developed countries who must have access to the Internet. This is where new knowledge is freely available.

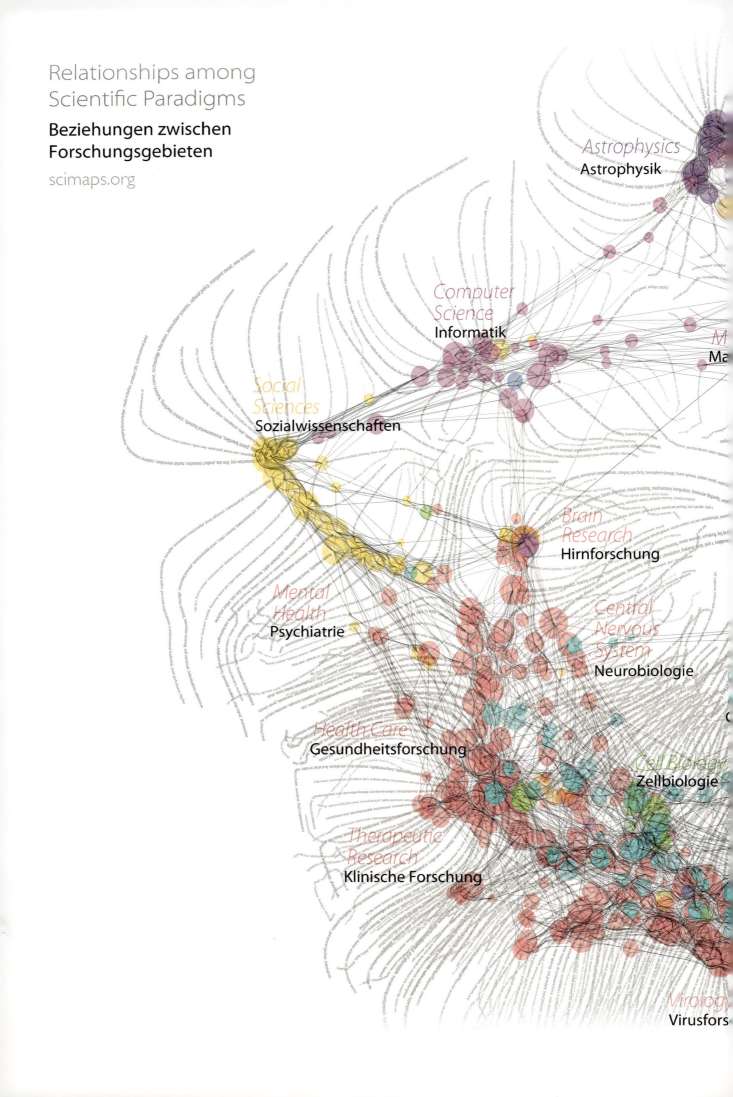

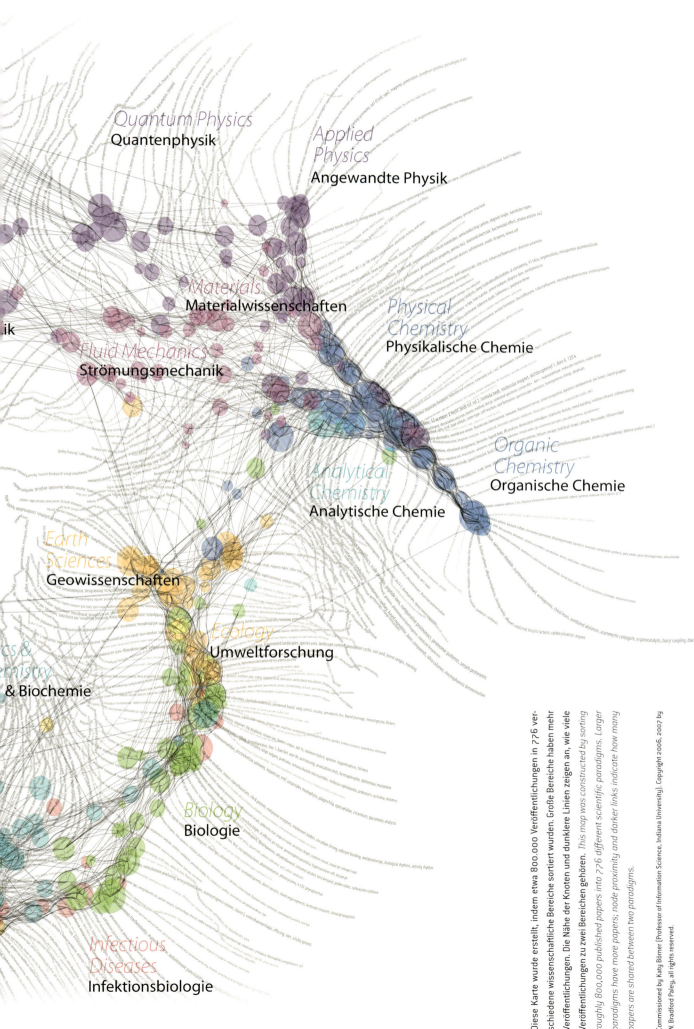

Is there a global brain in the making? *Globe-spanning networks of scientists and intelligent software help us to understand complex relationships. The digital networking of the many enhances the creativity of the individual – we live in ›accelerated times‹. Investments in research are rising worldwide. Measuring stations, scientists and knowledge are being networked together ever more intelligently; relationships between entities are being recognised earlier. We are witnesses to an explosion of human creativity. This transformation is driven by digital information and communication technologies. New knowledge is emerging all over the world in this day and age, and it will be available everywhere in the future. The decisive factor will then be who knows how to use it quickly, competently and adequately.* **Knowledge on demand** The Internet of the future will be your answering machine, personal assistant, advisor, extended memory and companion. Our relationship with knowledge is changing. The Internet is becoming a ›knowledge workshop‹: a constant supply of new tools are helping us to tap into diverse sources of knowledge, exchange experiences and find out about the latest zeitgeist just in time. **How do new qualities emerge?** High-performance computers and data networks are gradually enabling us to understand why the whole is often more than the sum of its parts. Analysis robots, ever stronger computational power, applied mathematics and fast data networks – we are gradually beginning to understand why the interaction of simple building blocks can produce brand new qualities or completely new behaviours. **Scientific freedom – the elixir of civilisation** Scientific and technical progress depends upon what freedom scientists have and how openly they are able to exchange ideas. The driving force of progress is the creativity of individuals. And this is growing in this day and age thanks to the global networking of scientists. As a result, projects get off the ground quicker, new ideas are taken up faster. **What does the digital future hold for us?** Goods, services and people are becoming digitally networked – blurring the boundaries between our work and home environments, between our real life and our virtual existence. Research is changing our lives. New research findings are being generated throughout the world in rapid succession. But the number of concrete applications does not rise automatically. Innovations must find their way into a highly-networked society that has become sensitive to change. The key factor is how open society is to new things and what needs and values people are driven by. New knowledge can lie idle or be taken up faster by others.

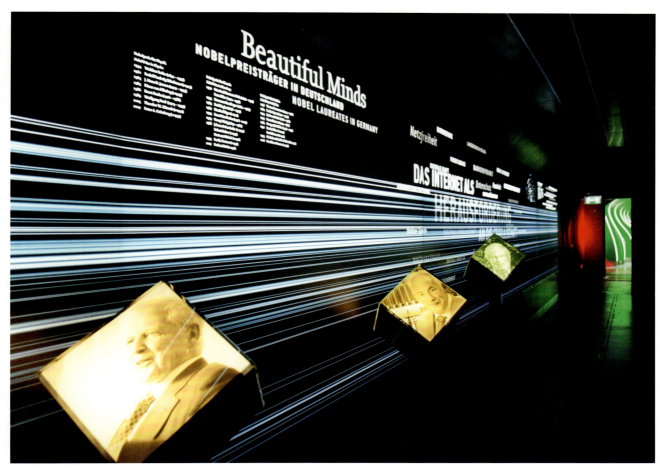

_____ **How will we live?** A new, global information and communication system is radically changing our thinking and our culture. In a gradual process, objects and services are becoming connected with each other through tiny entities. Everything is ›smart‹, ›intelligent‹ or ›virtual‹ – whether it's in the home, in production, in mobility and logistics or in healthcare provision and the energy sector. _____ **Where is the Internet heading?** The Internet of the future is about more than access, bandwidth and devices. Its potential lies in novel applications and digital content. The Internet of the future will bring information, goods, services and people together. The Semantic Web will help you find what you need. Networked systems will be with us at every turn. The objects and services around us will communicate with each other and with us.

1 Das Internet bietet neue Zugänge in die Welt des Wissens und neue Technologien schaffen neue Möglichkeiten darauf zuzugreifen. Jede der Karten auf dem Multimedia-Tisch enthält eine Webseite, von der sich die Besucher ausgewählte Inhalte auf den Monitoren ansehen können. *The Internet provides new access to the world of knowledge and new technologies create new ways to retrieve this. Each of the cards on the multi-media table contains a website, from which visitors can view selected content on the monitors.* _____ 2 Auch in einer vernetzten Welt, in der Forschungskollaborationen immer größer werden, kommt es am Ende auf die Ideen des Einzelnen an. *Even in a networked world with ever larger research collaborations it is the idea of the invidual that counts.*

NEUE MATERIALIEN UND DIE PRODUKTION DER ZUKUNFT
INNOVATIVE MATERIALS AND THE FACTORY OF THE FUTURE

intelligent + virtuell

1
2

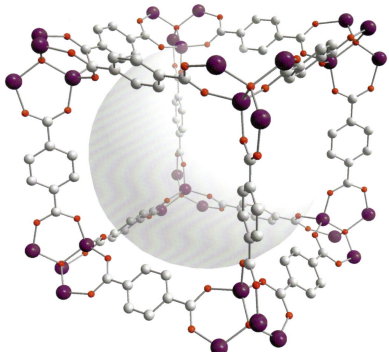

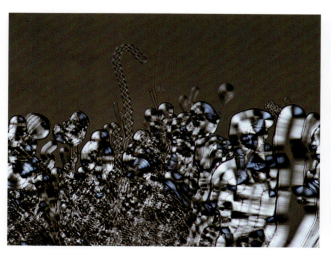

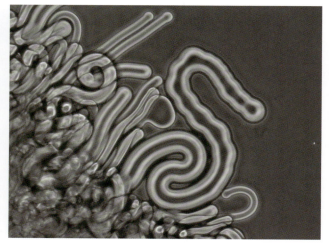

3
4

Neue Materialien und die Produktion der Zukunft

Künstliche Materialien werden kontinuierlich verbessert. Ihre neuartigen Eigenschaften tragen zusammen mit innovativen Fertigungsverfahren dazu bei, die Produktideen der Zukunft zu realisieren. Optimierte konventionelle sowie völlig neu entwickelte Materialien erfüllen maßgeschneidert vielfältige Funktionen. Häufig lassen sich die Forscher dabei von biologischen Prinzipien inspirieren. In den kommenden zehn bis zwanzig Jahren werden innovative Materialien viele Verfahren und Produkte revolutionieren, die uns im täglichen Leben begegnen. Die digitale Fabrik der Zukunft verknüpft alle Stationen des Lebenszyklus eines Produkts und ermöglicht so eine flexible Fertigung. **Werkstoffe und Technologien für die Zukunft** *Auch im 21. Jahrhundert sind zahlreiche Innovationen von gänzlich neuen Materialien und Technologien abhängig. Zudem bilden sie die Basis für moderne Produkte in allen Lebensbereichen. Das Verständnis des atomaren und molekularen Aufbaus der Materie schreitet rasant voran. Werkstoffe erhalten neuartige Eigenschaften. Materialien werden maßgeschneidert und vereinen verschiedene, manchmal sogar widersprüchliche Eigenschaften und Funktionen. Dies ermöglicht Anwendungen, die vorher undenkbar waren. Im Spannungsfeld zwischen Miniaturisierung, Funktionalität, Design und Nachhaltigkeit entstehen intelligente, individuelle und umweltfreundliche Produkte für die Zukunft.* **Grenzenlose Vielfalt** → **SUPRALEITEND** Wird die Vision von der verlustlosen Stromleitung bei Zimmertemperatur mit neu gefundenen Materialien Wirklichkeit? Seit kurzem begeistern eisenhaltige Pnictide die Forscher. Diese Hochtemperatursupraleiter verlieren ihren Widerstand bereits bei ›hohen‹ -220 °C. Sie besitzen eine einfache schichtartige Struktur und lassen sich sogar als Einkristalle züchten. → **MULTIFERROISCH** Kombiniert man ferromagnetische mit ferroelektrischen Eigenschaften, kann die Magnetisierung durch ein elektrisches Feld gesteuert werden, oder umgekehrt. Multiferroische Materialien führen zu neuartigen Bauelementen: Speicherbausteine werden energiearm mit Spannung statt mit Strom geschaltet. Diese ferroelektrischen Nanokondensatoren mit einer Speicherdichte nahe bei 150 GByte pro cm² sind einzeln ansteuerbar. → **WEICH** Tenside wie in Waschpulver oder Shampoo spielen eine Schlüsselrolle bei zahllosen chemischen Prozessen und Konsumgütern. Beim Auflösen in Wasser bilden Tenside selbständig hochgeordnete Gebilde. Dabei werden faszinierende Instabilitäten in Form von lamellenartigen Strukturen an der Grenzfläche zwischen Tensid und Wasser beobachtet. → **HOCHPORÖS** Die Größe der Poren metallorganischer Gerüstverbindungen lässt sich auf atomarer Skala genau einstellen. Diese eignen sich gut etwa für die Gasspeicherung. Gasatome finden ihren Platz in den winzigen Gerüstlöchern, die sich wie dreidimensionale Siebe verhalten. Die Gerüste haben ein großes Potenzial für die Gasreinigung, die Trennung von Stoffen und die chemische Katalyse. → **THERMOELEKTRISCH** Zwei Drittel der in technischen Prozessen eingesetzten Energie verpuffen als Wärme. Thermoelektrische Generatoren verwandeln diese Abwärme in Strom. Das Generatormodul beruht auf Würfeln von optimierten thermoelektrischen Halbleitermaterialien. Deren Ladungsträger liefern Strom bei einem Temperaturgefälle zwischen der Ober- und Unterseite. Heiße Autoabgase können so helfen, Energie zu sparen. **Ferromagnetische Flüssigkeiten** Ferrofluide sind Flüssigkeiten, die auf magnetische Felder reagieren. Ihre Oberfläche verformt sich dabei dreidimensional. Wasser oder Öl, das mit Nanopartikeln aus Eisen, Nickel oder Kobalt angereichert ist, lässt sich von Magneten steuern. Als Lacke verpassen Ferrofluide Flugzeugen eine Tarnkappe. Auch als Röntgenkontrastmittel und zur Tumorbehandlung sind sie geeignet. **Magnetismus maßgeschneidert** Materialien, deren magnetische Eigenschaften in alle drei Raumrichtungen angepasst werden können, besitzen hohes technologisches Potenzial. Die Entwicklung neuartiger magnetischer Materialien hat in

---

Vorige Doppelseite: Unterschiedliche Kristallstrukturen des keramischen Hochtemperatur-Supraleiters $YBa_2Cu_3O_{7-x}$. *Preceding spread: Different crystal structures of the ceramic high-temperature superconductor* $YBa_2Cu_3O_{7-x}$. **1** Nanowürfel aus metallorganischen Gerüstmaterialien. *Nanocubes of metal-organic frameworks.* **2** Hohlraum, der inmitten der Elementarzelle eines metallorganischen Gerüsts Gasmoleküle aufnehmen oder als Pore eines Siebs wirken kann. *Hollow space that can hold a gas molecule inside a unit cell of a metal-organic framework or that can work as a pore in a sieve.* **3, 4** Lamellenartige Doppelmembranen aus Tensid-Molekülen, die sich nach dem Kontakt mit Wasser (oben bzw. rechts) in faszinierende Strukturen umwandeln. *Lamellar double membranes of tenside molecules, which change to fascinating structures after contact with water (top, right).*

Vorige Doppelseite *Preceding spread* Ute Schäfer, Max-Planck-Institut für Metallforschung, Stuttgart **1** BASF SE **2** Prof. Dr. Stefan Kaskel, Institut für Anorganische Chemie, Technische Universität Dresden **3, 4** Prof. Dr. S. U. Egelhaaf, Institut für Physik der kondensierten Materie, Heinrich-Heine-Universität Düsseldorf

den vergangenen Jahren den gigantischen Anstieg der Speicherdichte von Computerfestplatten ermöglicht. Zukünftige Anwendungen liegen vor allem im Bereich der Sensorik und Nanoelektronik.⎯ **Unkonventionelle Lichtbrechung** Das Strukturieren von Materialien im Größenbereich der Lichtwellenlänge erlaubt es, deren optische Eigenschaften genau einzustellen. Optische Bauteile, die auf der Beugung des Lichts beruhen, sind kompakter und leichter zu handhaben als Linsen und Spiegel. Dies führt zu grundlegend neuen optischen Funktionen. Zudem ermöglichen Lithografieverfahren die billige Massenproduktion.⎯ **Elektronik aus Kunststoff** Elektronische Schaltungen aus leitfähigen Kunststoffen wird man bald auf flexible Folien drucken können: Chips aus dem Tintenstrahldrucker. Polymere spenden Licht: Das Herz organischer Leuchtdioden bilden Dünnschichttransistoren. In einer Matrix angeordnet formen sie flexible Displays. Diese können als elektronisches Papier, als Datenspeicher und in der Photovoltaik genutzt werden.⎯ **Kunststoffe mit Intelligenz** Die Kunststoffe der Zukunft sind leicht, langlebig und wieder verwertbar. Sie lassen sich gezielt maßschneidern und besitzen erstaunliche Funktionen. Form und Größe von Polymerbauteilen können durch äußere Stimuli wie die Temperatur gezielt gesteuert werden. Das Schalten zwischen im ›Gedächtnis‹ fest eingeprägten und abrufbaren Formen bereichert die minimalinvasive Chirurgie und Montagetechnik.⎯ **Design neuer Materialien** Neue Materialien werden am Computer konzipiert. Geistesblitze liefert auch direkt die Natur, die viele technische Fragen bereits elegant gelöst hat. Ideen aus der Tier- und Pflanzenwelt führen zu innovativen Lösungsansätzen. Immer leistungsfähigere Simulationstechniken sagen Eigenschaften und Verhalten von Werkstoffen zuverlässig voraus. Dies erspart oft experimentelle Tests bei der Entwicklung.⎯ **Revolution in der Produktion** *Die Rahmenbedingungen und*

1 Mikrohärchen mit Pilzkopf (rechts) bilden das Geheimnis eines neuen Haftmaterials, dessen Oberflächenstruktur von den Käferfußsohlen (links) inspiriert ist. *Micro-hairs with mushroom-shaped ends (right) are the secret of a new adhesive material with a surface structure inspired by the soles of the beetle's feet (left).* 2 Ferromagnetische Flüssigkeit: Ferrofluide sind Flüssigkeiten, die auf magnetische Felder reagieren. Ihre Oberfläche verformt sich dabei dreidimensional. *Ferromagnetic liquid: Ferrofluids are liquids that react to magnetic fields. Their surface deforms three-dimensionally in the process.*

1 PD Dr. Stanislav N. Gorb, Evolutionary Biomaterials Group, Max-Planck-Institut für Metallforschung, Stuttgart 2 IBM Deutschland GmbH. Copyright (c) 2001 Sachiko Kodama, Minako Takeno. All Rights Reserved.

Möglichkeiten industrieller Produktion wandeln sich ständig durch neue Technologien, Materialien oder Organisationsformen. Daraus entstehen neuartige Anforderungen an die Planung und Realisierung von Fabriken. Die Konkurrenzfähigkeit von Unternehmen hängt von ihrer Fähigkeit ab, kundenspezifische Produkte höchster Qualität herstellen zu können, die zudem technologisch führend sind. Um solche Produkte erfolgreich zu entwickeln, setzen viele Firmen Methoden des Innovationsmanagements ein. Doch meist fehlen den Betrieben die optimalen Werkzeuge, um eine dazu passende flexible und leistungsfähige Produktion zu planen. Die heutige Massenproduktion steht vor großen Herausforderungen. **Neuartige Produktionstechniken** 3D-Siebdruck: Die Industrie hat das klassische Druckverfahren weiter entwickelt, um komplexe dreidimensionale keramische und metallische Objekte herzustellen. Durch Siebdruck hergestellte Hohlkammersysteme erreichen maximale Stabilität bei niedrigstem Gewicht. Solche, durch konventionelle Verfahren bisher nicht produzierbaren Elemente eignen sich als Dieselpartikelfilter und Wärmetauscher. **Die Digitale Fabrik** Als ganzheitliche Planungsumgebung bildet die Digitale Fabrik Produkte und Fertigungsprozesse ab. Dazu wird leistungsfähige Software benötigt. Durch dynamische digitale Nachbildungen, Materialfluss-Simulationen, ergonomische Untersuchungen, Kapazitätsplanung oder Zeitanalysen können Produktionsstätten realitätsnah modelliert und die Abläufe in Echtzeit simuliert werden. **Von der Idee zum marktfähigen Produkt** Die schnelle Herstellung von präzisen Musterbauteilen ist für Unternehmen eine wichtige Voraussetzung, konkurrenzfähig innovative Produkte zu entwickeln. Beim Rapid Prototyping werden die Teile erst am Computer entworfen und dann direkt an 3D-Drucker exportiert. Diese bauen die Objekte aus Kunststoff, Metallstaub oder Porzellan auf. Die Materialpalette erweitert sich ständig, auch die Genauigkeit nimmt zu. **... und wie geht es weiter?** Der zunehmende Zeit- und Kostendruck im globalen Wettbewerb macht es für Unternehmen unerlässlich, Produktionsprozesse kontinuierlich zu optimieren. Forschung und Entwicklung arbeiten an innovativen Lösungen für Produktionstechnologien und -verfahren. Flexibilität, Nachhaltigkeit, Integration von Produkt und Dienstleistung, Kooperationsmanagement und Nanotechnologie markieren wichtige Trends.

Zahlreiche Dinge, die in unserem Alltag eine Rolle spielen, werden automatisiert hergestellt. Diese Produktionsstraße demonstriert die vollautomatische Herstellung von Fußbällen. Numerous things that play a role in our day-to-day lives are manufactured through automation. This fully automated production line demonstrates how footballs are made.

Leihgeber *On loan by* Siemens AG

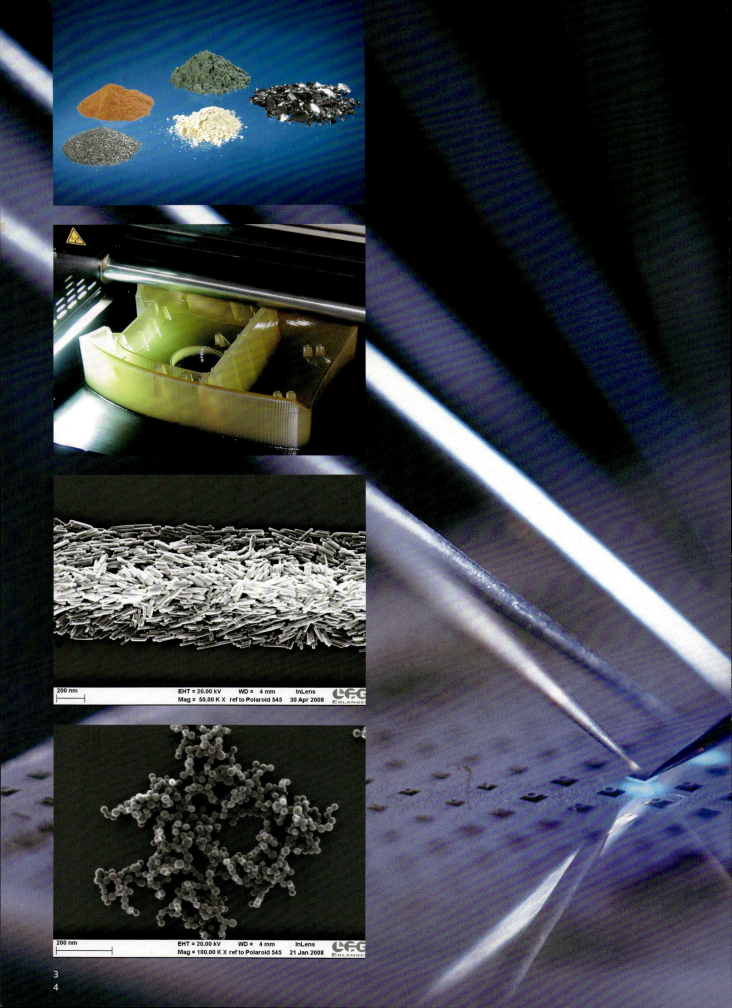

Innovative materials and the factory of the future

Synthetic materials are continually being improved. Their novel properties combined with innovative production techniques help to develop the new product ideas of the future. Optimised conventional materials and brand new materials alike are tailor-made to fulfil a diverse range of functions. These days, the researchers that develop them are often inspired by biological principles. In the coming ten to twenty years, innovative materials will revolutionise many of the techniques and products we encounter in our daily lives. The digital factory of the future will link all stages in the product lifecycle, thereby enabling flexible manufacturing.

Materials and technologies for the future *Even in the 21st century, many innovations depend on completely new kinds of materials and technologies. These also form the basis of modern products in all spheres of life. Our understanding of the atomic and molecular structure of matter is growing at a breathtaking pace. Substances are being imbued with new qualities. Materials are being customised, uniting varied and sometimes even conflicting properties and functions. This enables previously unimaginable applications. Between the competing priorities of miniaturisation, functionality, design and sustainability we are seeing the evolution of intelligent, individual and environmentally-friendly products for the future.*

Boundless variety → **SUPERCONDUCTING** Will the vision of zero-loss current conduction at room temperature one day be a reality thanks to newly invented materials? Scientists recently started to get excited about ferrous pnictides. The high-temperature superconductors lose their resistance at a ›high‹ -220 °C. They have a simple layered structure and can even be grown as monocrystals. → **MULTIFERROIC** If ferromagnetic properties are combined with ferroelectric qualities, the magnetisation can be controlled by an electric field, or vice versa. Multiferroic materials give rise to new components: Memory chips can be switched on a low-energy basis with voltage instead of current. These ferroelectric nanocapacitors can be addressed individually and have a storage density of nearly 150 GByte/cm^2. → **SOFT** Surfactants, like those found in washing powder or shampoo, play a crucial role in countless chemical processes and consumer goods. When dissolved in water, surfactants independently take on a highly ordered form. Fascinating instabilities can be observed in the form of lamellar structures at the interface between the surfactant and the water. → **HIGHLY POROUS** The size of the pores in metal-organic frameworks can be set precisely in the atomic scale. Such frameworks are suitable for gas storage, for example. Gas atoms find their place in the tiny pores, which act like three-dimensional sieves. The frameworks hold considerable potential for gas cleaning, the separation of materials and chemical catalysis. → **THERMOELECTRIC** Two thirds of the energy used in technical processes escape in the form of heat. Thermoelectric generators convert this heat into electricity. The generator module is based on cubes of optimised thermoelectric semiconducting materials. Their charge carriers deliver electricity at a temperature gradient between the upper and the lower side. Hot exhaust fumes from cars can thus help to save energy.

Ferromagnetic liquids Ferrofluids are liquids that react to magnetic fields. Their surface deforms three-dimensionally in the process. Water or oil enriched with nanoparticles of iron, nickel or cobalt can be controlled by magnets. Used as paints, ferrofluids can camouflage aircraft. They can also be used as an X-ray contrast medium and to treat tumours.

Magnetism tailor-made Materials with magnetic properties that can be adapted in all three dimensions possess substantial technological potential. The development of innovative magnetic materials was what enabled the relatively recent, enormous rise in the storage density of computer hard drives. Future applications lie primarily in the fields of sensor technology and nanoelectronics.

Unconventional refraction Materials structured on the scale of light wavelengths allow their optical

---

1 Alle Metalle und Legierungen, die als Pulver verfügbar sind, etwa Eisen, Kupfer, Nickel, Wolfram oder Stähle, können zu beliebigen Formen gedruckt werden. *All metals and alloys available in powder form, such as iron, copper, nickel, tungsten and steel, can be printed to produce any shape.* 2 Ein kompliziertes Bauteil aus Kunststoff entsteht im 3D-Drucker wie von Geisterhand. *A complicated component made of plastic is created by a 3D printer as if by magic.* 3 Dieser Nanodraht ist aus Halbleiteroxid aufgebaut. Der Halbleiter gehört zu den elektronischen Materialien der Zukunft. *This nano wire made of semiconductor oxide is made up of tiny rods. The semiconductor is one of the electronic materials of the future.* 4 Ziel ist es, auch konventionelle Halbleiterelektronik zu drucken. Diese Silizium-Nanokugeln, in Schichten angeordnet, könnten die Bausteine dafür bilden. *The goal is to be able to print even conventional semiconductor electronics. These silicon nanospheres, arranged in layers, could provide the building blocks for the technology.*

1 Fraunhofer-Institut für Fertigungstechnik und Angewandte Materialforschung (IFAM), Institutsteil Pulvermetallurgie und Verbundwerkstoffe, Dresden 2 RTC Rapid Technologies GmbH 3, 4 DFG-Exzellenzcluster »Engineering of Advanced Materials«, Erlangen *Hintergrund Background* Fraunhofer IAF

properties to be precisely set. Optical components based on the diffraction of light are more compact and easier to handle than lenses and mirrors. This yields fundamentally new optical functions. Lithography techniques also enable them to be mass-produced inexpensively. **Electronics from plastic** It will soon be possible to print electronic circuits made of conductive plastics onto flexible foils: chips from an inkjet printer. Polymers provide light: thin-film transistors are at the heart of organic light-emitting diodes. Arranged in a matrix structure, they form flexible displays. These can be used as electronic paper, as a data storage medium and in photovoltaics. **Intelligent plastics** The plastics of the future will be light, durable and recyclable. They can be customised for the task at hand and possess amazing functions. The shape and size of polymer components can be controlled by external stimuli, such as temperature. Switching between ›memorised‹ and retrievable shapes enhances minimally-invasive surgery and installation engineering. **Designing new materials** New materials are designed on the computer. Nature itself provides some of the inspiration, having already found elegant solutions for many technical issues. Ideas from the animal and plant kingdom are leading to innovative potential solutions. Ever more powerful simulation techniques are reliably predicting the properties and performance of materials. This often saves on experimental tests in the development phase.
Revolution in production *The underlying conditions and possibilities in industrial production are constantly changing as new technologies, materials or forms of organisation develop. This creates new requirements for the planning and realisation of factories. A company's competitiveness depends on its ability to produce customer-specific products of the highest quality that are also technologically state-of-the-art. In a bid to develop such products successfully, many companies employ innovation management techniques. Usually, however, companies lack the optimal tools to suitably plan flexible and efficient production process. Modern-day mass production faces major challenges.*
Innovative production techniques 3D screen printing: The industry has worked to evolve the traditional printing method into a technique capable of manufacturing complex, three-dimensional ceramic and metallic objects. Hollow chamber systems manufactured using the screen printing method achieve maximum stability with the lowest-possible weight. Such elements, still unachievable using conventional methods, can be used as diesel particle filters or heat exchangers. **The Digital Factory** As an integrated planning environment, the Digital Factory pictures products and production processes. Powerful software is needed to achieve this. Using dynamic digital mock-ups, material flow simulations, ergonomic tests, capacity planning or time analyses, production facilities can be realistically modelled and the processes simulated in real time. **From concept to marketable product** The ability to manufacture precise prototypes quickly is one of the keys to developing innovative products competitively. Rapid prototyping is a process in which the parts are first designed on the computer and then exported straight to 3D printers. The printers construct the objects out of plastic, metal dust or porcelain. The range of materials is growing all the time, as is the precision of the technique. **... and what comes next?** Faced with growing pressure on time and costs in an atmosphere of global competition, companies have no choice but to continually optimise their production processes. Research and development teams are working on innovative solutions for production technologies and processes. Flexibility, sustainability, integration of product and service, cooperation management and nanotechnology characterise the important trends.

Das Ziel der Digitalen Fabrik ist die ganzheitliche Planung, Evaluierung und laufende Verbesserung aller wesentlichen Strukturen, Prozesse und Ressourcen der realen Fabrik in Verbindung mit dem Produkt. *The objective in the Digital Factory is a holistic planning: Evaluation and continous improvement of all important structures, processes and resources in the real factory in combination with the product.*

Fraunhofer IWU, Chemnitz

EINE KLEINE WELT FÜR SICH
A SMALL WORLD OF ITS OWN

hinter den kulissen

Mind-map

Veränderung Alltag

Ernährung

- pflanzliche ausgestoßene Arzneimittel
- mehr künstliche Zusatzstoffe

Arbeit

- Die Arbeit wird, wie heute, immer mehr durch Roboter ersetzt, viele Menschen werden arbeitslos.
- Mehr Arbeitslose, weil Maschinen Arbeit übernehmen
- Roboter die die Arbeit ersetzen = mehr Arbeitslosigkeit
- mehr zu hause arbeiten
- Mehr Arbeitsplätze
- mehr Jobs und besser bezahlt

Medizin

- künstliche Organe
- Heilmittel für schlimme Krankheiten

Fortbewegung

- Wasser-Autos
- ~~Wasserstoff~~ Autos Wasserstoff-Autos
- Umweltfreundlicher
- Hybridautos Fahrräder
- Autos und wie, Wasserstoff, umweltfreundlich

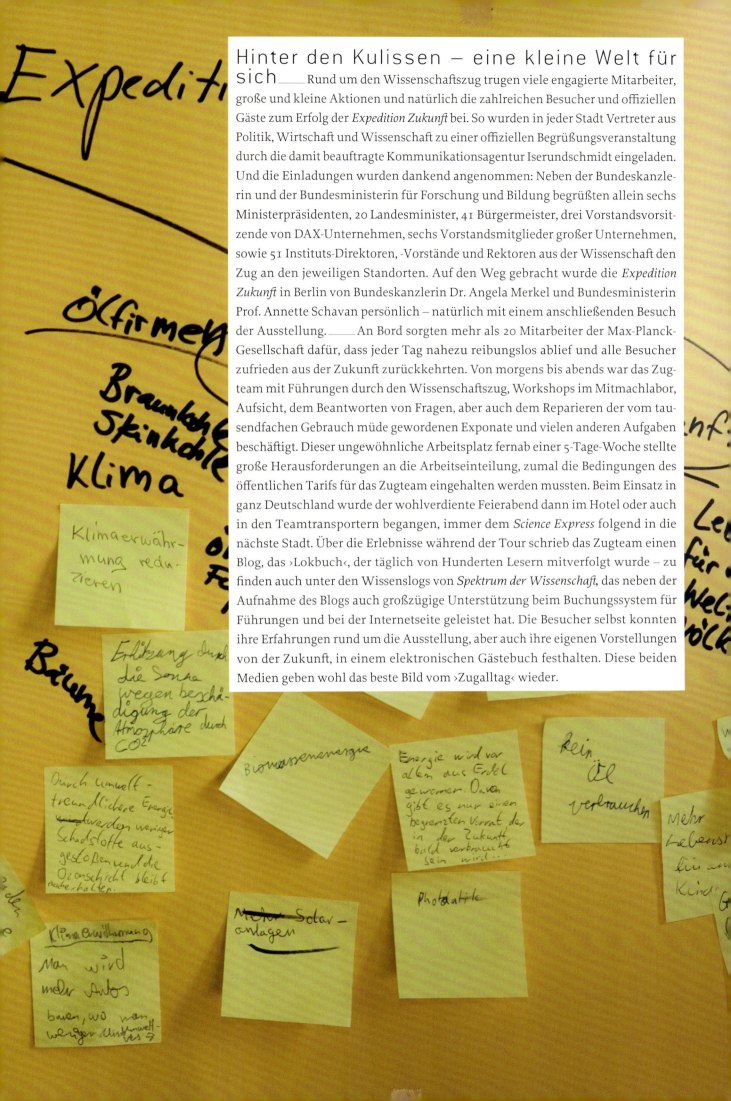

Hinter den Kulissen – eine kleine Welt für sich

Rund um den Wissenschaftszug trugen viele engagierte Mitarbeiter, große und kleine Aktionen und natürlich die zahlreichen Besucher und offiziellen Gäste zum Erfolg der *Expedition Zukunft* bei. So wurden in jeder Stadt Vertreter aus Politik, Wirtschaft und Wissenschaft zu einer offiziellen Begrüßungsveranstaltung durch die damit beauftragte Kommunikationsagentur Iserundschmidt eingeladen. Und die Einladungen wurden dankend angenommen: Neben der Bundeskanzlerin und der Bundesministerin für Forschung und Bildung begrüßten allein sechs Ministerpräsidenten, 20 Landesminister, 41 Bürgermeister, drei Vorstandsvorsitzende von DAX-Unternehmen, sechs Vorstandsmitglieder großer Unternehmen, sowie 51 Instituts-Direktoren, -Vorstände und Rektoren aus der Wissenschaft den Zug an den jeweiligen Standorten. Auf den Weg gebracht wurde die *Expedition Zukunft* in Berlin von Bundeskanzlerin Dr. Angela Merkel und Bundesministerin Prof. Annette Schavan persönlich – natürlich mit einem anschließenden Besuch der Ausstellung._____ An Bord sorgten mehr als 20 Mitarbeiter der Max-Planck-Gesellschaft dafür, dass jeder Tag nahezu reibungslos ablief und alle Besucher zufrieden aus der Zukunft zurückkehrten. Von morgens bis abends war das Zugteam mit Führungen durch den Wissenschaftszug, Workshops im Mitmachlabor, Aufsicht, dem Beantworten von Fragen, aber auch dem Reparieren der vom tausendfachen Gebrauch müde gewordenen Exponate und vielen anderen Aufgaben beschäftigt. Dieser ungewöhnliche Arbeitsplatz fernab einer 5-Tage-Woche stellte große Herausforderungen an die Arbeitseinteilung, zumal die Bedingungen des öffentlichen Tarifs für das Zugteam eingehalten werden mussten. Beim Einsatz in ganz Deutschland wurde der wohlverdiente Feierabend dann im Hotel oder auch in den Teamtransportern begangen, immer dem *Science Express* folgend in die nächste Stadt. Über die Erlebnisse während der Tour schrieb das Zugteam einen Blog, das ›Lokbuch‹, der täglich von Hunderten Lesern mitverfolgt wurde – zu finden auch unter den Wissenslogs von *Spektrum der Wissenschaft*, das neben der Aufnahme des Blogs auch großzügige Unterstützung beim Buchungssystem für Führungen und bei der Internetseite geleistet hat. Die Besucher selbst konnten ihre Erfahrungen rund um die Ausstellung, aber auch ihre eigenen Vorstellungen von der Zukunft, in einem elektronischen Gästebuch festhalten. Diese beiden Medien geben wohl das beste Bild vom ›Zugalltag‹ wieder.

Behind the scenes — a small world of its own

In and around the science train, the hard-working staff, many major and minor activities and, of course, the numerous visitors and official guests all contributed to the success of the *Science Express*. In every city, representatives from politics, business and science were invited to an official welcoming event by the project's PR agency, Iserundschmidt. The invitations were gratefully accepted: In addition to the German Federal Chancellor and the Federal Minister of Education and Research six Prime Ministers of German federal states, 20 state ministers, 41 mayors, three CEOs of DAX companies, six board members of large companies, as well as 51 directors, university rectors and executive committee members from the science sector greeted the train in various locations. The *Science Express* was launched in Berlin by German Federal Chancellor Dr. Angela Merkel and Prof. Annette Schavan, the Federal Minister of Education and Research – followed, as a matter of course, by a tour of the exhibition. ⎯ Aboard, more than 20 employees of the Max Planck Society ensured that the exhibition ran practically without a hitch, and that all visitors emerged highly satisfied from their excursion into the future. From morning to evening, the train team was busy with tours of the exhibition, workshops in the hands-on laboratory, supervising visitors, answering questions, repairing the exhibits that had suffered from the strong wear and tear arising from their handling by thousands of visitors, and many other tasks. This unusual place of work, very different from the normal five-day week, presented huge challenges in relation to the division of tasks, as the working conditions applicable to public service employees had to be complied with. During the tour throughout Germany, the team's well-earned evenings were spent in hotels or on the team transporters hot on the trail of the *Science Express* to the next city. The train team wrote a blog about its experiences during the tour, the ›Lokbuch‹ (›train's log‹), which was followed by hundreds of readers. The blog could also be found among the Wissenslogs, the German ›SciLogs‹ on the website of *Spektrum der Wissenschaft*, the prominent German scientific journal, which, in addition to publishing the blog, also provided generous support for the exhibition's tour booking system and the project website. Visitors could also record their experiences and impressions of the exhibition as well as their ideas about the future in an electronic guest book. These provide the best account of ›everyday train life‹.

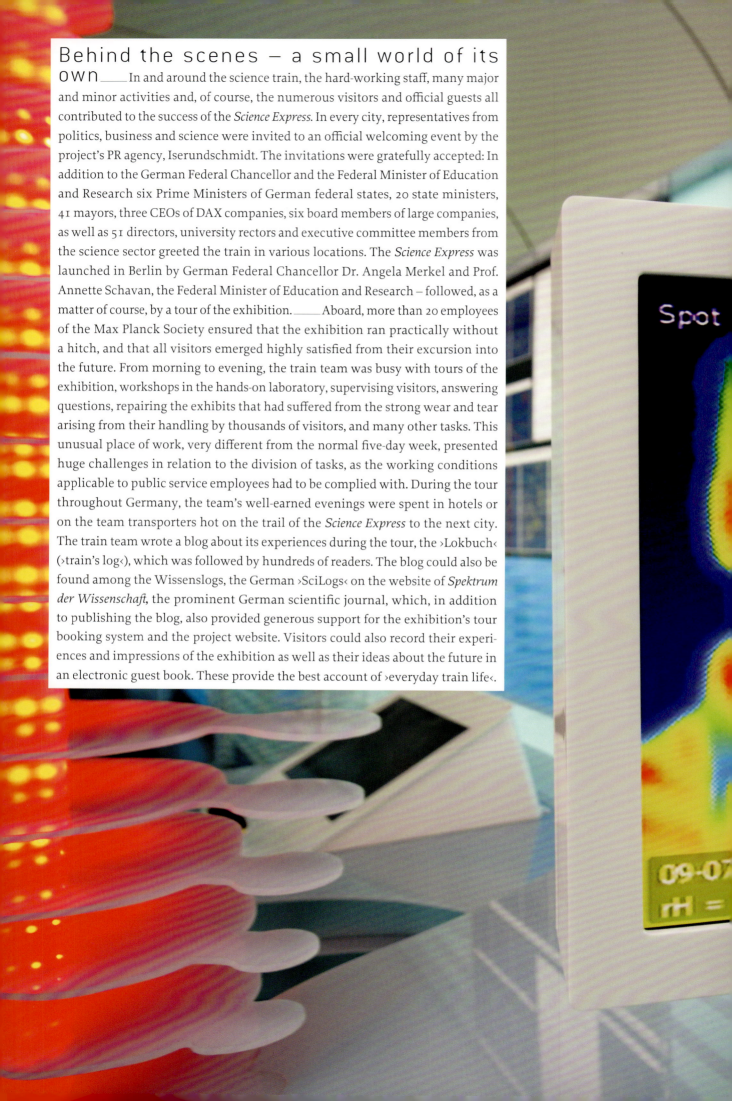

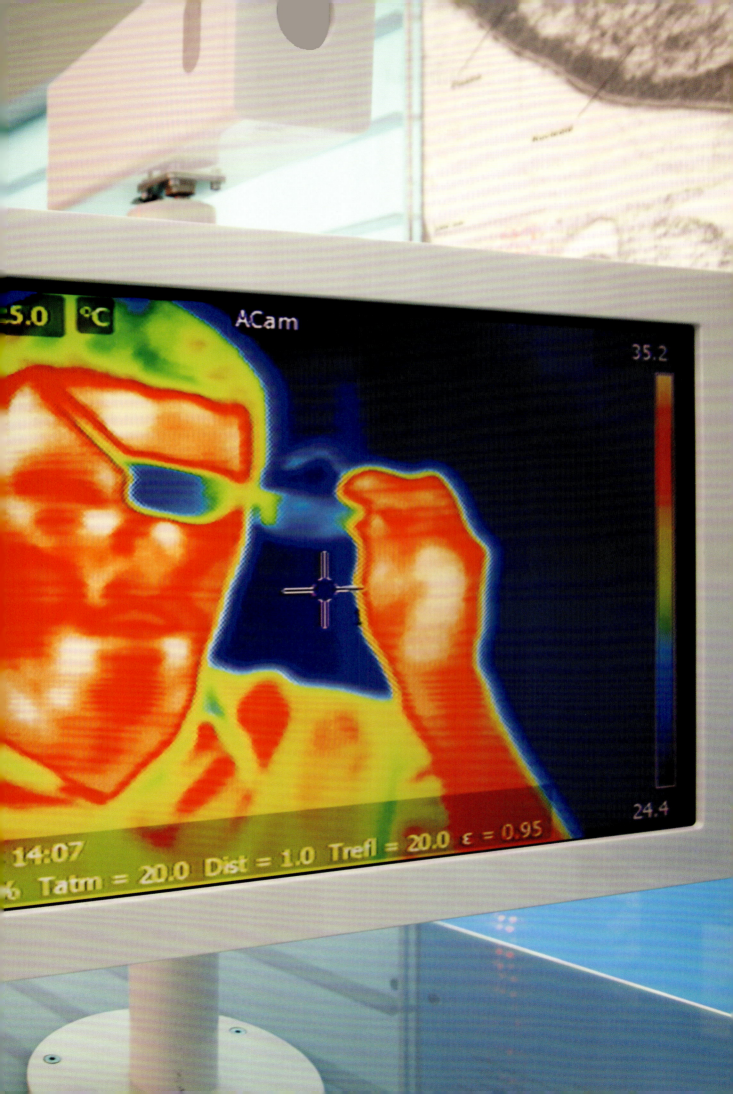

Gute Fahrt und viel Freude bei der "Reise" mit Zug "Expedition Zukunft"

DR. ANGELA MERKEL
Bundeskanzlerin der Bundesrepublik Deutschland

Lernen und Forschen gehört zu den besten Seiten des Menschen! Das steht auch diesem Zug: allzeit gute Fahrt!

PROF. DR. ANNETTE SCHAVAN, MDB
Bundesministerin für Bildung und Forschung

Ich wünsche dem Zug eine gute Reise und viele begeisterte Besucher!

PROF. DR. PETER GRUSS
Präsident der Max-Planck-Gesellschaft zur Förderung der Wissenschaften e.V.

Viel Erfolg mit dem Zug in die Zukunft

PROF. WAN GANG
Minister of Science and Technology of the People's Republic of China

ERÖFFFNUNG DER AUSSTELLUNG
23. APRIL 2009
BERLIN

EXPEDITIONZUKUNFT
SCIENCEXPRESS
WISSENSCHAFTSAUSSTELLUNG IN MEHR ALS 60 STÄDTEN
WWW.EXPEDITION-ZUKUNFT.ORG

Forschungs-expedition Deutschland

Es war sehr interessant – wir als Schülerinnen konnten probieren/ausprobieren und durften pleuten. Danke.

Jetzt weiß ich endlich, wer meinen Hausputz macht: mein Besen alleine! Danke für die Idee.

WAR GUT BZW. IRRE!!!

Ich wünsche all denen, die mit unschätzbarem Einsatz und Wissen all das, was hier dokumentiert wird, ermöglicht haben und weiter entwickeln, dass sie sich immer mehr unseres Menschseins bewusst bleiben. Möge bei aller Verantwortung die Liebe ständiger Begleiter sein.
(Ursula, 65 Jahre)

I found the idea good, but it was a bit too long... but my brother was completely spellbound!

Eine tolle Konzeption für Klein und Groß!
Wir hatten viel Spaß und haben viel gelernt.
Anja + Karl mit Simon
+ Moritz
Bonn, 31.05.2009

Super team, friendly supervision – thank you!
Excellent lecture, very clear and not at all dry :)
Interesting experiments :)
Max-Ernst-Gymnasium
Cologne, 28.05.2009

This was the best day of my life.
Leverkusen, 9.6.2009

Good, funky!

Muchas gracias, muy interesante el recorrido, bastante información.

That thing was really awesome!

That was an amazing experience! What a pity that you were only here for three days!!! I hope that this kind of event will be repeated some time!!!

Nina hat sich über die hoppelnden Bürsten fast totgelacht! Vielen Dank!

Die Schülings
Jülich, 04.06.2009

WIR SIND BEGEISTERT VON DIESEN SCIENCE EXPRESS!
WIR DANKEN IHNEN FÜR DIE ZAHLREICHEN INFORMATIONEN
UND WÜNSCHEN IHNEN VIEL ERFOLG UND GUTE REISE
MIT FREUNDLICHEN GRÜSSEN ZWEI SCHÜLER DES
MARIE-CURIE GYMNASIUMS DRESDEN

Die Wissenschaft ist der Wahnsinn!
Der Zug ist der Hammer und es sollte noch mehr zu sehen geben!
Vielen Dank, Schmitt

Echt krass das Teil!

THAT WAS A WONDERFUL EXPERIENCE. THERE WAS SO MUCH TO LEARN. MY BRAIN IS BURSTING NOW!

Stimmen aus dem Blog **Donnerstag, 23. April, 13 Uhr** Auf Gleis 2 im Berliner Hauptbahnhof herrscht angespannte Stille. Kein Zug fährt ein oder aus, keine Passagiere warten ungeduldig auf dem Bahnsteig. Nur die Anzeigentafel kündigt das kommende Ereignis an: Die rollende Ausstellung *Expedition Zukunft* fährt zum ersten Mal in einen Bahnhof ein. **Frankfurt, 25.–27. April** Lachende Gesichter sehen wir auch bei vielen anderen Besuchern, die bei dem schönen Wetter in Frankfurt den Zug als Ziel für einen Wochenendausflug eingeplant haben. ›So einen Zug müsste es öfter geben‹ und ›ein riesiges Lob, einfach toll‹ hören wir natürlich gerne. **Ludwigshafen, 26.–28. Juni** Einige Kinder warten ungeduldig darauf, dass das Mitmachlabor seine Pforten öffnet. Schließlich nähern sie sich der magischen Uhrzeit 15:30 Uhr. Der Countdown läuft. Doch nicht nur gedacht. Nein, die Kinder zählen laut mit: ›Fünf … Vier … Drei … Zwei … Eins … Null!!!‹ Prompt öffnet sich die Glastür und strahlende Kindergesichter werden eingelassen. Wer sagt denn, dass Silvester der schönste Countdown des Jahres sein muss. **Karlsruhe, 2.–4. Juli** Außerdem gibt es natürlich auch in Karlsruhe die ein oder andere nette Szene mit unseren Besuchern. So geht ein Vater mit seinem Sohn durch Wagen 3, bio + nano. Dort steht ein 3D-Bildschirm, an dem man organische Moleküle betrachten kann. Der Vater stellt fest: ›Dieses 3D, das ist schon schwierig, da macht das Gehirn irgendwann nicht mehr mit.‹ Daraufhin der Sohn, ganz trocken: ›Meins schon.‹ **Tübingen, 22.–23. Juli** Zwei Tage haben die Tübinger Zeit, den Zug aufzusuchen – und die ca. 80.000 Einwohner zählende Stadt mit ungefähr 24.000 Studenten stürmt am zweiten Tag geradezu den Zug: Trotz Regenschauer werden beinahe 2.000 Besucher gezählt! **Erfurt, 9.–11. August** Direkt am Sonntagmorgen warten so viele Besucher vor dem Zug, dass wir von Anfang an eine Schlange haben. Ein Anblick, der Jörgs poetische Ader anspricht: ›Was für eine lange Schlange‹, sagt er mit bedeutungsschwangerer Stimme, ›mir wird bange.‹ Diszipliniert, wie die Erfurter sind, behalten sie die Schlange im Zug bei. Auch im ersten Themenwagen stehen sie brav in Zweierreihen an. Irgendwann traut sich ein Besucher zu fragen: ›Dürfen wir auch überholen?‹ Aber sicher! Wir bitten darum! **Görlitz, 16.–18. August** Verwundert konnte ich einen älteren, großen, schlanken Herrn beobachten, der wie ein Storch im Salat vor der Wärmebildkamera herumstakste. Seine Erklärung: Beim Hochheben seiner Beine wollte er testen, ob er sein künstliches Kniegelenk auf dem Bildschirm erkennen könne – und tatsächlich, es wurde in kaltem Blau wiedergegeben. **Straubing, 23. September** Auf dem Weg von Bamberg nach Bayreuth lassen wir uns Straubing nicht entgehen. Und wenn wir schon mal da sind, dann nutzen die Straubinger das auch aus. An einem Tag kommen so viele Besucher, wie an anderen Standorten an drei Tagen zusammen. Morgens haben wir noch nicht einmal geöffnet und trotzdem schon die ersten 200 Besucher im Zug. Es standen nämlich so viele Schüler vor dem Eingang, dass wir sie nicht warten lassen wollten … So viele Schulklassen an einem Tag hat der Zug wahrscheinlich noch nie gesehen. **Aschaffenburg, 27.–29. September** Der *Science Express* sorgt bei einer Aschaffenburger Mutter für wahre Existenzängste. Gefangen im Dschungel der Wissenschaft, kämpft sie sich ihrem Sohn hinterher, nur um am Ende festzustellen: Ihr Sohn hat noch eine weitere Etappe für sich entdeckt. Von jedweder Hoffnung verlassen, ruft sie verzweifelt: ›Oh nein! Er ist im Mitmachlabor gelandet! Wir kommen hier nie wieder raus.‹ Das muss ein wahrer Forscher sein, der die Welt um sich herum vergisst (Mutter eingeschlossen) und sich begeistert seinem Projekt widmet. Unsere Bildungsmission bei diesem Jungen ist erfüllt.

Voices from the Blog

Thursday, April 23rd, 1 pm An expectant hush prevails on Platform 2 of Berlin Hauptbahnhof railway station. There is no train arriving or leaving, no passengers waiting impatiently on the platform. The forthcoming event is announced only by the information display: the rolling exhibition *Science Express* is about to pull into a train station for the first time.

Frankfurt, April 25–27 We also see smiling faces among many other visitors who have made the train their destination for a weekend excursion in the lovely weather in Frankfurt. Naturally we are delighted to hear such comments as: ›This kind of train should happen more often‹ and ›very well done, simply brilliant‹.

Ludwigshafen, June 26–28 Some children are waiting impatiently for the hands-on laboratory to open its doors. The magic time, 3.30 pm, finally approaches. The countdown is on. Not just in their heads. No, the children count out loud: ›Five ... four ... three ... two ... one ... zero!!!‹ The glass door opens promptly and the children, their faces beaming, are allowed in. Who says that the best count down of the year is on New Year's Eve.

Karlsruhe, July 2–4 Needless to say, there are a few nice scenes with our visitors in Karlsruhe as well. A father and his son go through carriage 3, *bio + nano*. There is a 3D monitor there on which you can see organic molecules. The father comments: ›This 3D thing is really difficult, at some point your brain just can't keep up anymore.‹ To which his son answers dryly: ›Mine can.‹

Tübingen, July 22–23 The people of Tübingen have two days to visit the train – and the city, which has around 80,000 inhabitants and approximately 24,000 students practically storms the train on the second day: despite the rain, almost 2,000 visitors are recorded!

Erfurt, August 9–11 On Sunday morning there are so many visitors waiting in front of the train that we have a queue from the outset. A sight that appeals to Jörg's poetic sensibility: ›What a long queue,‹ he says in a portentous voice, ›What shall we do?‹ The very disciplined people of Erfurt keep the queue up in the train. Also in the first carriage they stand well-behaved in rows of two. Eventually one visitor dares to ask: ›May we overtake?‹ Of course! Please do!

Görlitz, August 16–18 I was amazed to see a tall, slim elderly gentleman manoeuvring awkwardly in front of the thermal imaging camera. His explanation: he wanted to test whether he could see his artificial knee joint on the screen by lifting up his leg. And, lo and behold, it was actually there to see in the cold-blue.

Straubing, September 23rd En route from Bamberg to Bayreuth we don't want to miss out on Straubing. And now that we are here, the people of Straubing do take advantage of our train. In one day we have so many visitors as on three days in other locations. In the early morning, while the exhibition is not yet open, there are already the first 200 visitors on board. There were so many pupils in front of the entrance that we did not want to keep them waiting ... The train probably hasn't seen so many school classes in one day ever before.

Aschaffenburg, September 27–29 The *Science Express* causes existential fear in a mother from Aschaffenburg. Ensnared in the science jungle she struggles after her son only to learn at the end: her sun has discovered a further stage. Deserted by hope, she despairingly exclaims: ›Oh no! He found the learning lab! Now we'll never get out of here.‹ This has to be a researcher in the true sense of the word, who forgets all the world around himself (including his mother) and dedicates himself wholly to his project. With him our mission has been achieved.

Der Zug und das Bordpersonal wiederum wurden rund um die Uhr betreut von jenem Projektteam in der Generalverwaltung der Max-Planck-Gesellschaft in München, das zuvor die gesamte Ausstellung konzipiert, entwickelt, organisiert und gemeinsam mit der Agentur ArchiMeDes umgesetzt hatte. Hotels und Aushilfskräfte, Führungen und Workshops wurden organisiert, Informationsmaterialien nachgeliefert, Anfragen bearbeitet und Webseiten gepflegt. Quer durch die Republik reisten die Mitarbeiter zum Wissenschaftszug, mal als Zugchef, um die offiziellen Gäste mit auf die *Expedition Zukunft* zu nehmen, mal als Aufsicht, als Pressesprecher oder auch als Kaffeeboy, um die Stimmung im Team hoch zu halten. ──── Und konnte ein defektes Exponat einmal nicht durch die technischen Mitarbeiter vor Ort repariert werden, so waren die Ausstellungsbauer von ArchiMeDes spätestens zum regelmäßigen Wartungstag zur Stelle, um den Schaden zu beheben. Zum Gelingen haben ebenso die Planer und Umsetzer des Mitmachlabors, two4science, beigetragen wie die Imago GmbH bei der Organisation von Sonderführungen in den Abendstunden. Doch das größte Lob gebührt den Besuchern, ihrem Interesse, ihrer Geduld und ihrem Engagement. So griffen die vielen Rädchen zuverlässig und mit hoher Präzision ineinander und ergaben einen großartig funktionierenden Mechanismus und für alle Beteiligten eine schöne Zeit.

The train and the staff on board, in turn, were supported round the clock by the project team at the Administrative Headquarters of the Max Planck Society in Munich, which had conceived, developed, coordinated and jointly implemented the entire exhibition with the ArchiMeDes agency. Hotels and temporary support staff, tours and workshops were organised, information material replenished, enquiries processed and websites maintained. Staff from the Headquarters also travelled across the whole of Germany to the train, to act as train manager and to welcome official guests aboard, help out with supervision, to work as press officers and even to get the coffee and generally keep the team's spirit up. ──── If a broken exhibit could not be repaired by the technical staff on site, the ArchiMeDes exhibition planners were there to deal with it – at the latest on the regular maintenance days. two4science, who planned and implemented the hands-on laboratory also made an invaluable contribution to the success of the venture, as did Imago GmbH who helped with the organisation of special evening tours. But the highest praise belongs to the visitors, in their interest, patience and commitment. So the many gears interlocked reliably and with maximum precision, and together produced an amazing mechanism – it was a wonderful time for all those involved.

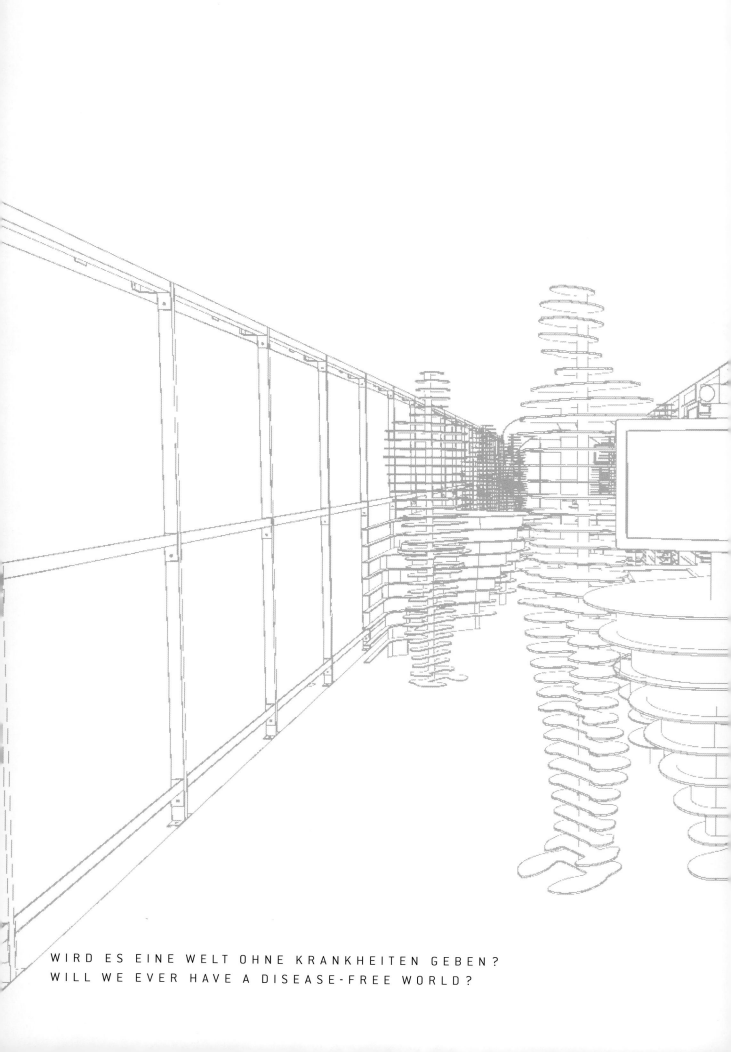

WIRD ES EINE WELT OHNE KRANKHEITEN GEBEN?
WILL WE EVER HAVE A DISEASE-FREE WORLD?

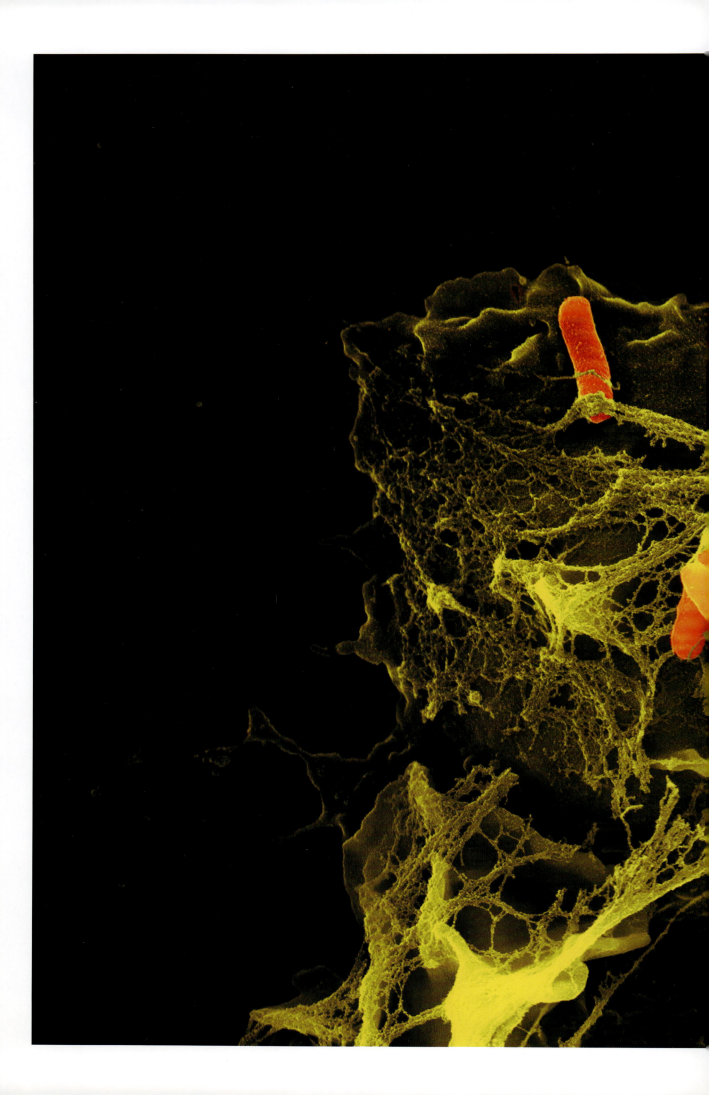

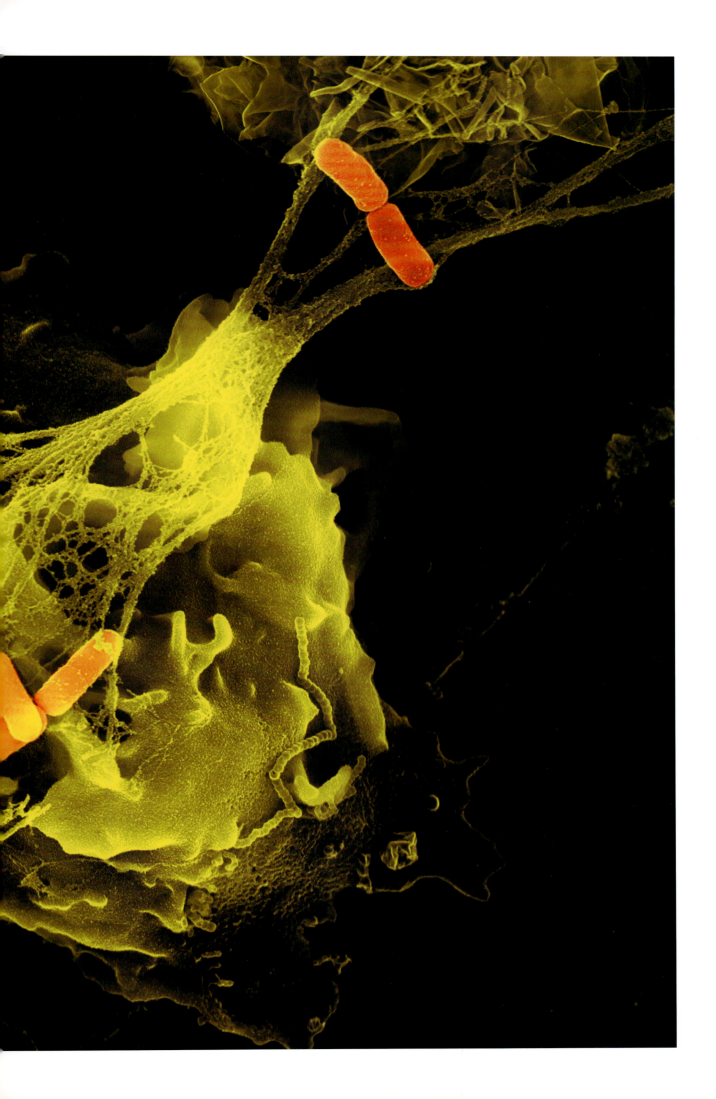

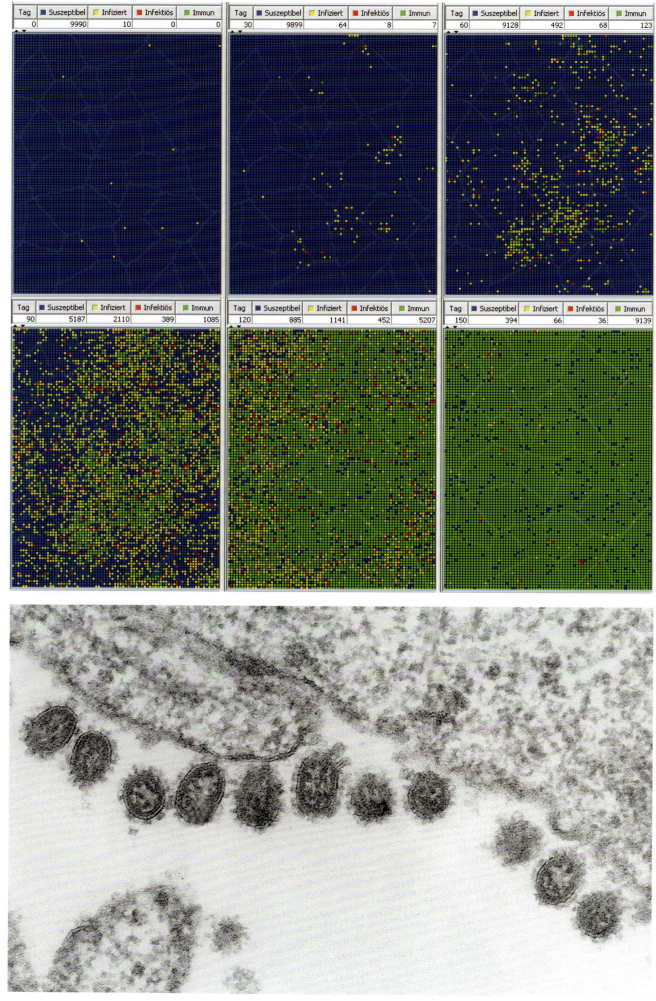

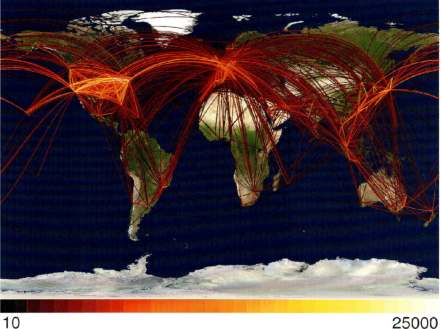

10   25000

Wird es eine Welt ohne Krankheiten geben?

Die Lebenswissenschaften werden die Gesellschaft im 21. Jahrhundert maßgeblich prägen. Fundamentale neue Erkenntnisse eröffnen bislang ungeahnte Möglichkeiten Krankheitsauslöser frühzeitig aufzuspüren und neue Therapien zu entwickeln. Die Genomforschung ermöglicht Innovationen in der präventiven und individualisierten Medizin, die neue Dimensionen der Vorbeugung und medizinischen Versorgung versprechen. Dank intensiver Forschung wächst das Spektrum von Technologien zur klinischen Diagnostik und für operative Eingriffe sowie zur Herstellung individueller Ersatzgewebe und -organe rasch. Neben all diesen Chancen müssen wir auch lernen, mit dem zukünftigen Potenzial der Medizin verantwortungsvoll und gerecht umzugehen. **Die globalen Krankheiten unserer Zeit** *Die Ursachen zahlreicher Krankheiten sind inzwischen bekannt. Dennoch steht unsere Gesundheit nach wie vor Bedrohungen gegenüber, die wegen ihrer Komplexität oder Wandelbarkeit Jahr für Jahr Millionen Menschenleben fordern. Trotz enormer Fortschritte in der Medizin bleiben viele, weit verbreitete Krankheiten unheilbar und einige davon nehmen sogar epidemische Ausmaße an. Dazu gehören Infektionskrankheiten wie Grippe und Aids, aber auch chronische und komplexe Erkrankungen wie Diabetes und Herz-Lungen-Leiden. Depressionen und Alzheimer nehmen in der Bevölkerung aufgrund des rasanten Lebenswandels und der Altersstruktur zu. Auch Krebs ist nach wie vor eine der häufigsten Todesursachen weltweit.* **Erreger jetten um die Welt** In einer global vernetzten Welt haben Krankheitserreger leichtes Spiel. Ein Virus könnte so leicht eine Pandemie, eine weltweite Epidemie, auslösen. Um Deutschland auf eine Pandemie vorzubereiten, werden nicht nur Impfstoffe entwickelt und Überwachungsmaßnahmen getroffen. Auch Computersimulationen, die alle Eventualitäten berücksichtigen, tragen dazu bei, einer drohenden Katastrophe vorzubeugen. **Woran erkranken wir?** Trotz medizinischer Fortschritte sind einige Krankheiten noch immer unberechenbar und Ursache der meisten Invaliditäts- und Todesfälle in Deutschland. An den Ursachen der weit verbreiteten Krankheiten wie beispielsweise Herz-Kreislauf-Störungen, Demenz oder Krebs wird intensiv geforscht. Könnten diese eines Tages geheilt werden, würden wir eine völlig neue Lebensqualität erreichen. **Ein neues Zeitalter der Medizin bricht an** *Die eigenen Gene bergen das Geheimnis, welche Krankheiten jeden einzelnen in Zukunft erwarten. Durch die Entschlüsselung des menschlichen*

Vorige Doppelseite: In Netzen, gebildet von speziellen Leukozyten, verfangen sich unerwünschte Eindringlinge wie etwa Bakterien. *Preceding spread: Unwanted intruders such as bacteria get caught in a mesh formed by special leukocytes.* ▬▬ 1 Der Epidemiesimulator EpiDyNet zeigt die Ausbreitung von Infektionen in einer Bevölkerung. Jede Person ist durch ein kleines Quadrat dargestellt. Ohne Eingriff würde ein Pockenausbruch fast die gesamte Bevölkerung erreichen. *The epidemic simulator EpiDyNet shows the spread of infection in a population. Each person is represented by a small square. Without intervention, a smallpox outbreak would infect almost the entire population.* ▬▬ 2 Influenza – mit dem Auslöser der echten Virusgrippe infizieren sich jährlich 5 bis 15 Prozent der Weltbevölkerung, 250.000 bis 500.000 davon sterben daran. *Influenza – between 5 and 15 percent of the world's population is infected each year by the viral flu; between 250,000 and 500,000 people die from it.* ▬▬ 3 Erreger jetten um die Welt: Die Farbigkeit der Linien symbolisiert die Intensität des Flugverkehrs zwischen den 500 größten Flughäfen weltweit. *Pathogens jet around the world: The colors of the lines indicate the intensity of air traffic between the 500 biggest airports.*

Vorige Doppelseite *Preceding spread* Volker Brinkmann, Max-Planck-Institut für Infektionsbiologie, Berlin 1 Dr. Markus Schwehm, ExploSYS GmbH, Leinfelden, www.explosys.de 2 Robert Koch Institut, Berlin 3 Max-Planck-Institut für Dynamik und Selbstorganisation, Göttingen

Genoms werden immer mehr Genmutationen entdeckt und Methoden entwickelt, unser Erbgut direkt zu behandeln. An vielen Krankheiten wie Krebs, Asthma oder Parkinson sind auch mutierte Gene beteiligt. Durch die Identifizierung der daran beteiligten, verdächtigen DNS-Abschnitte versuchen die Forscher nun, Krankheiten direkt an ihrem Ursprung zu behandeln. Dazu werden neue Verfahren entwickelt, um diese Gene aufzufinden, zu manipulieren oder sogar aus dem Erbgut herauszuschneiden. Auf umgekehrtem Weg können ›gesunde‹ Gene in die Zelle gebracht werden – doch nicht nur zu therapeutischen Zwecken. **DNS-Spürhund für Gendefekte** *Die Diagnose der Zukunft verlangt einen Tropfen Blut und einen DNS-Chip. Auf Krankheitsgene programmiert, findet dieser individuelle Defekte im Erbgut. Bevor ein Chip zum Einsatz kommt, muss das defekte Gen für eine Krankheit erst gefunden werden. Beim erblichen Autismus ist dies bereits geglückt. An Versuchstieren werden die Erkenntnisse überprüft, bevor eine Therapie für den Mensch entwickelt wird.* **Molekulare Schere gegen Aids** *Eine HIV-Infektion lässt sich nicht heilen. Daher versuchen Forscher nun, die tödliche Krankheit durch Behandlung der Erbsubstanz zu besiegen. Bei einer HIV-Infektion baut das Virus sein Erbgut in das menschliche Genom ein. Statt die Viren aus dem Körper zu entfernen, ist es Forschern erstmals gelungen, das fremde Erbgut mit einem speziellen Enzym aus der DNS des Menschen auszuschneiden.* **Das Wirken der Gene verändern** *Ähnlich Maschinen in einer Fabrik werden auch die biologischen Maschinen in einer Zelle nach Bedarf reguliert. Können wir bald am Kontrollpult stehen? Immer besser verstehen wir es, von außen einzelne Gene an- und auszuschalten oder gezielt zu therapieren. Doch auch am gesunden Menschen ist der Eingriff in das Erbgut möglich: Es könnte manipuliert werden, um etwa unsere Leistungsfähigkeit zu steigern.* **Ingenieure werden die neuen Halbgötter in Weiß** *In der Medizin von morgen spielt der Einsatz moderner Technologien eine große Rolle. Nicht nur für den Arzt werden Diagnose und Therapie durch hochentwickelte Geräte und Verfahren immer effektiver. Auch der Patient profitiert von geringeren Belastungen. Die Medizintechnik der Zukunft begrenzt sich nicht auf innovative Apparate, die einzelne Krebszellen im Körper erkennen können oder dem Chirurgen die Bilder über den Operationsfortschritt live auf einem Display wiedergeben. Wenn Sie eines Tages einen Zahnersatz brauchen oder ein neues Herz, könnten medizinisch-biologisch gefertigte Gewebe oder intelligente Prothesen mit modernster Technik dazu beitragen, wieder gesund und munter in die Welt lächeln zu können.* **Dem Krebs einen Schritt voraus** *Krebs verursacht rund ein Viertel aller Todesfälle. Je früher die Krankheit erkannt wird, desto besser die Heilungschancen. Daher wird besonders an neuen Methoden der Früherkennung geforscht. Könnten bereits die ersten Krebszellen im Körper durch Biomarker aufgespürt und mit modernen Geräten lokalisiert werden, wären belastende Operationen oder Chemotherapien nicht mehr nötig.* **Ersatzteile aus eigenen Zellen** *Ist es zukünftig möglich, dass wir selbst zu unserer Heilung beitragen? Intensiv wird daran geforscht, aus eigenen Zellen Ersatzgewebe zu züchten. Bioingenieure versuchen beim ›tissue engineering‹, bei der künstlichen Erzeugung von körpereigenen Geweben, menschliche Zellen so umzuprogrammieren, dass aus ihnen neue Gewebe heranwachsen. Diese könnten ohne Abstoßungsreaktionen transplantiert werden.* **Intelligente Prothesen retten Leben** *Ein gesundes Herz ist ein Segen, ein krankes lebensgefährlich. Arbeitet ein Organ nicht mehr, könnte in Zukunft eine Prothese seine Funktion übernehmen. Intelligente Prothesen erfüllen komplexe Aufgaben im menschlichen Organsystem. Einige ›Ersatzteile‹ kommunizieren sogar mit der Außenwelt und sind somit cleverer als das Original. Welches Organ meldet heute schon im Voraus, dass es repariert werden muss?* **Medikamente von übermorgen** *Fortschritte in der Genomsequenzierung, Proteinchemie und Bioinformatik helfen bei der innovativen Ent-*

1 Maligne Lymphome: Pro 100.000 Einwohner gibt es jedes Jahr 15 Neuerkrankungen. *Malignant lymphoma: Each year there are 15 new incidences for every 100,000 inhabitants.* 2 Auswertung des DNS-Chips: Farben geben Auskunft über die Aktivität jedes Gens. Krebszellen und gesunde Zellen unterscheiden sich im Aktivitätsmuster. *Evaluation of the DNA chip: The colours provide information about the activity of each gene. The activity patterns of cancer cells and healthy cells are different.* 3 Neue Wirkstoff-Kandidaten durchlaufen umfassende Labortests, ehe sie mit Tieren und später mit Menschen erprobt werden können. *New candidate agents go through comprehensive laboratory testing before they can be tested on animals and then on humans.*

1 Kompetenznetze Medizin, gefördert vom BMBF, www.kompetenznetze-medizin.de 2 Deutsches Krebsforschungszentrum Heidelberg 3 Bayer AG

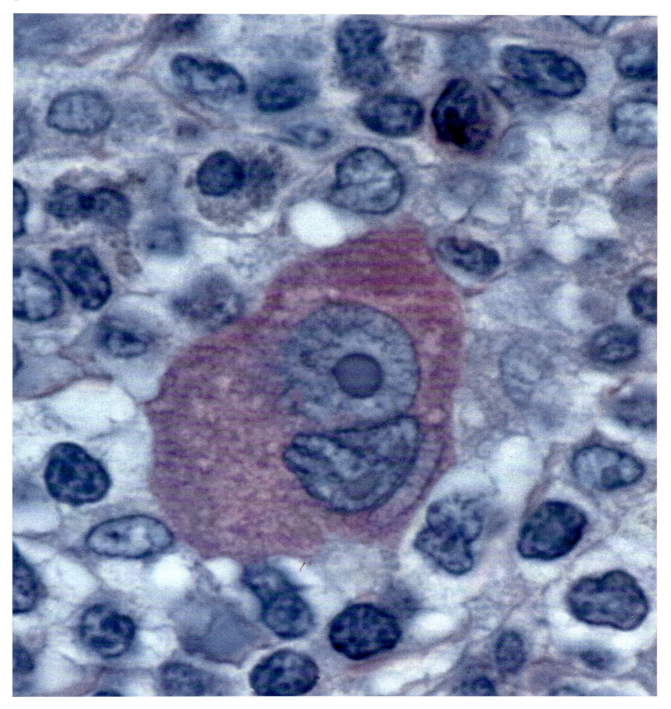
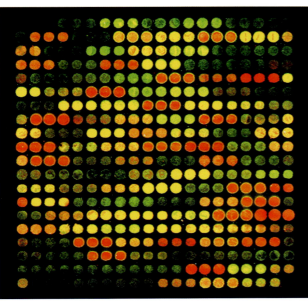

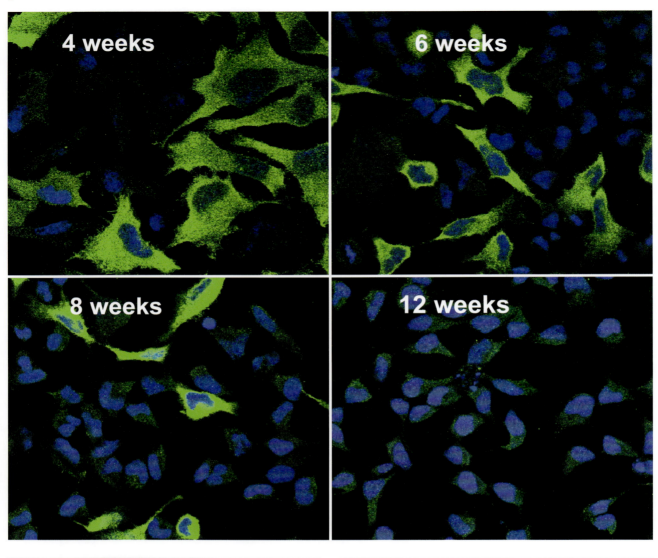

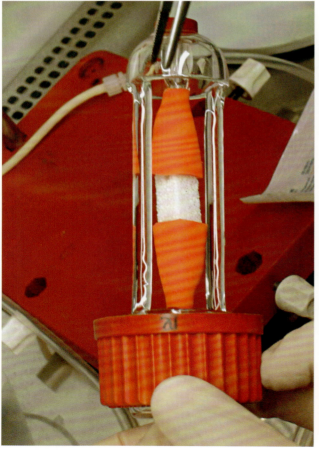

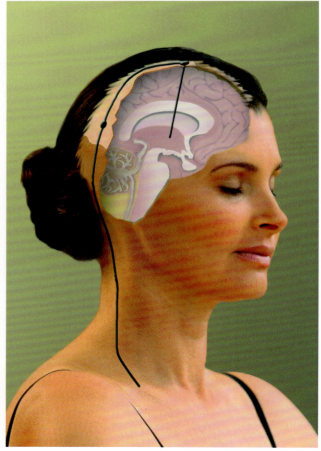

wicklung neuer Medikamente. Mit enormem technologischem Aufwand wird nach neuen chemischen Substanzen für Medikamente gefahndet, die tödliche Krankheiten bekämpfen. Oftmals sind Chemiker bereits in der Lage, passende Wirkstoffe am Computer zu entwickeln und maßgeschneidert herzustellen.

Operationstechniken 2020 Mithilfe neuer Entwicklungen in Endoskopie und Nanotechnologie wird an Methoden geforscht, umfassende chirurgische Eingriffe zu ersetzen. So könnten zukünftig magnetische Nanocontainer, gesteuert durch Magnetfelder, einen Wirkstoff direkt zum Krankheitsherd bringen oder endoskopische Kapseln krankhafte Zellveränderungen einfach aus dem Gewebe herausschneiden.

Der Operationssaal der Zukunft Im Hybrid-OP wird der Chirurg von einem innovativen Röntgensystem unterstützt. Durch die ständige Kontrolle verbessert sich die Sicherheit für Arzt und Patient. Die digitale Röntgenkamera, ein für Röntgenstrahlen durchlässiger OP-Tisch: Fortwährend kann der Arzt den Fortschritt der Operation kontrollieren. Dabei kommt die Operation mit Mini-Schnitten und einem Herzkatheter aus – zum Vorteil des Patienten.

Die Patientenakte wird digital Mit wenigen Klicks können alle Beteiligten im Gesundheitswesen wichtige Patientendaten für Diagnose und Therapie sicher untereinander austauschen. In der elektronischen Patientenakte sind alle wichtigen Daten hinterlegt und mit Zugriffsrechten geschützt – eHealth-Lösungen vermeiden Doppeluntersuchungen und erhöhen die Effizienz, weil sie den Informationsfluss verbessern und beschleunigen.

1 Das gezielte Design einer molekularen Schere (Rekombinase) erlaubt es, das HI-Virus aus kultivierten Zellen herauszuschneiden (grün: infizierte Zellen, blau: nicht infizierte Zellen). *Thanks to the tailored design of molecular scissors (recombinase), it is possible to cut the HI-virus out of cultivated cells (green: infected cells, blue: not infected cells).* 2 Der Bioreaktor fördert die Reifung des zellbesiedelten mineralischen Trägermaterials, das zur Rekonstruktion von Kieferdefekten implantiert wird. *A bioreactor promotes maturation of a cell-seeded mineral scaffold toward bone like tissue for reconstruction of jaw bone defects by implantation.* 3 Tief im Gehirn verursachen fehlgeschaltete Nervenzellen Symptome wie Tremor oder Muskelstarre. Die Elektroden des Hirnschrittmachers helfen, die Ursachen der Symptome abzuschalten. *Defective nerve cells deep inside the brain trigger symptoms such as tremors and muscular rigidity. The electrodes in the brain pacemaker help to block the signals that cause the symptoms.* 4 Besucher können die Funktionsweise einer myoelektrisch gesteuerten Armprothese testen. Über die Haut abgenommene Muskelsignale steuern die Prothesenbewegung. *Visitors can test the function of an armprothesis controlled by myoelectrodes. The prothesis is operated by muscle signals detected on the surface of the skin.*

1 Dr. Frank Buchholz, Max-Planck-Institut für Molekulare Zellbiologie und Genetik, Dresden 2 Translationszentrum für Regenerative Medizin (TRM), Jan Liese, Leipzig 3 Forschungszentrum Jülich 4 Leihgeber *On loan by* Otto Bock

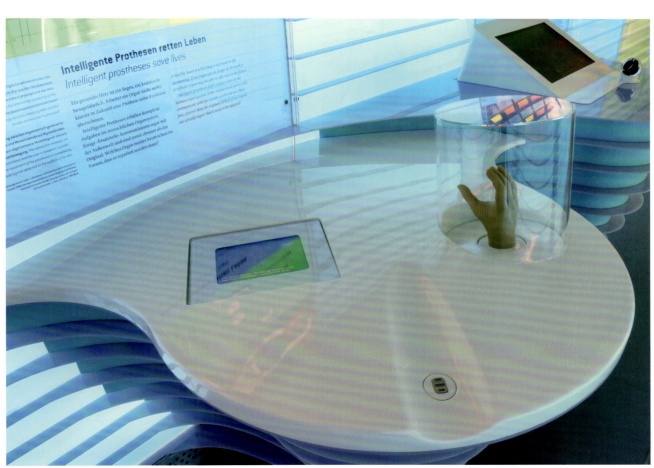

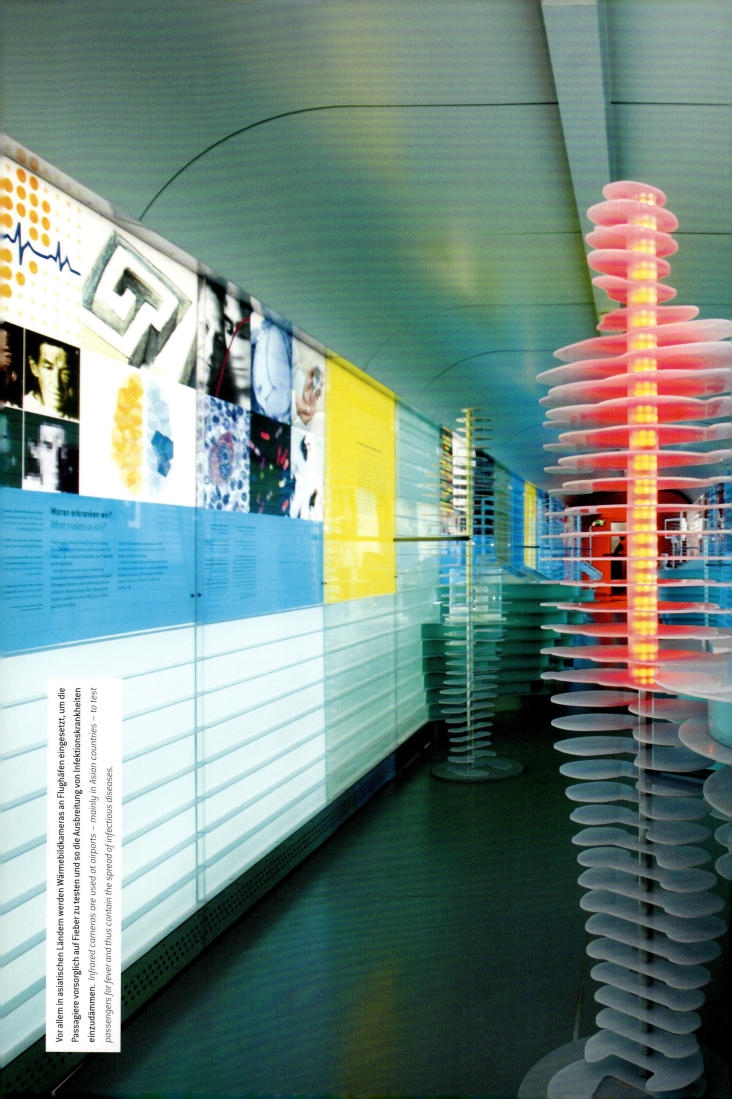

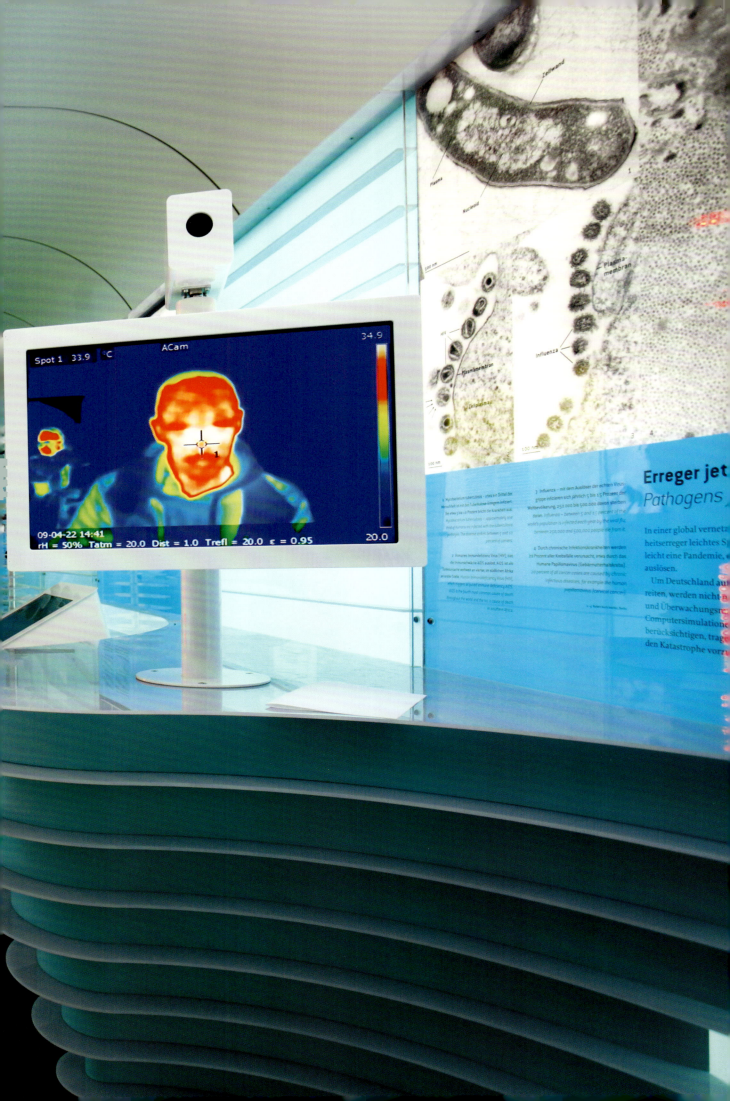

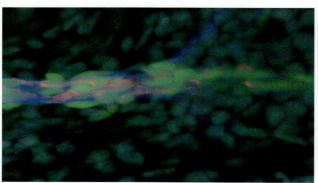

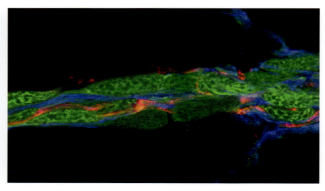

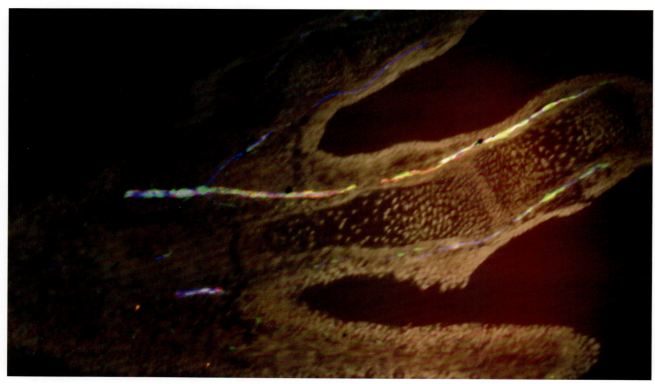

Will we ever have a disease-free world?

The life sciences will exert a crucial influence on society this century. Fundamental new insights will provide previously unthinkable options for the early identification of the factors that trigger disease and for the development of new treatments. Genome research enables innovations in preventive and individualised medicine, which promise new departures in relation to prevention and healthcare. Thanks to intensive research, the range of technologies available for clinical diagnostics and surgical interventions and for the production of replacement tissue and organs is growing rapidly. In addition to exploiting all of these new opportunities, we must also learn how to make responsible and fair use of the future potential of medicine.

The global diseases of our time We now know the causes of numerous diseases. However, our health continues to be at risk from various threats, which, due to their complexity and mutability, claim millions of lives, year after year. Despite enormous progress in medicine many widespread diseases remain incurable, some even assuming epidemic proportions. These include infectious diseases like influenza and Aids, as well as chronic and complex illnesses like diabetes and cardiovascular disorders. Due to fast-paced lifestyles and the changing age structure, the incidence of depression and Alzheimer's disease is increasing. Cancer also remains one of the main causes of death throughout the world.

Pathogens jet around the world Pathogens have it easy in today's small world. As a result, it would be very easy for a virus to trigger a pandemic, a global epidemic. Germany's preparation for a possible pandemic does not stop at the development of vaccinations and implementation of monitoring measures. Computer simulations that take every eventuality into account also contribute to the prevention of a disaster of this kind.

What makes us sick? Despite medical progress, some diseases remain unpredictable and are the cause of most cases of disability and death in Germany. Intensive research is being carried out on the causes of widespread illnesses such as cardiovascular disease, dementia and cancer. If we manage to cure these diseases one day, we will attain a whole new quality of life.

A new age of medicine is dawning One's own genes carry the secret as to which diseases await them in the future. Thanks to the decoding of the human genome, more and more genetic mutations are being discovered and methods developed for the direct treatment of our genetic material. Mutated genes are a contributing factor in many diseases such as cancer, asthma and Parkinson's. Through the identification of the DNA sections that are suspected as playing a role here, researchers are now trying to treat the diseases directly, at their origin. New procedures are being developed for locating disease genes, manipulating them and even cutting them out of the genetic material. Conversely, ›healthy‹ genes can be introduced into the cells – not just for therapeutic purposes.

The DNA ›sniffer dog‹ for genetic defects In future, all that will be required to make a diagnosis is a drop of blood and a DNA chip. These chips, which are programmed to locate disease genes, can find individual defects in the genotype. The defective gene for a disease must be found before a chip can be used. This has already been achieved in the case of hereditary autism. The findings are tested on laboratory animals before a treatment is developed for humans.

Molecular scissors against Aids HIV infection cannot be cured. Thus, researchers are now trying to conquer this fatal disease through the treatment of the genotype. When HIV infection occurs, the virus inserts its DNA into the human genome. Instead of removing the viruses from the body, researchers have succeeded for the first time in cutting out the alien DNA from the human genome with the help of a special enzyme.

Altering the way genes work Like machines in a factory, the biological machines in a cell are adjusted as required. How soon will we be able to take over the controls? We are increasingly better able to understand how to switch individual genes on and off, and to administer targeted treatment to them. However, it is also possible to intervene in the genome of healthy people: for example, it could be manipulated to increase our ability to perform. ——— **Engineers are the new saviours in white coats** *The use of modern technologies will play a significant role in the medicine of tomorrow. Thanks to highly developed devices and processes, diagnosis and therapy are becoming more and more effective. Doctors are not the only beneficiaries here; patients also gain through the reduction in suffering. The medical technology of the future is not limited to innovative devices that can identify cancer cells in the body or transmit images of surgical progress to the surgeon live on a display. Should you need a dental prosthesis or even a new heart some day, biomedically engineered tissues or intelligent prostheses based on the very latest technology may ensure that you will be able to keep smiling.* ——— **One step ahead of cancer** Cancer is responsible for around one quarter of all deaths. The earlier the disease is diagnosed, the greater the chances of recovery. For this reason, research is being carried out on new methods of early cancer detection in particular. If the first cancer cells can be detected in the body through biomarkers and localised using modern devices, stressful operations and chemotherapy would no longer be necessary. ——— **Replacement parts generated from your own cells** Will it be possible for us to contribute to our own healing and cure in the future? Intensive research is being carried out on a generation of substitute tissue from a patient's own cells. With tissue engineering, bioengineers are trying to reprogram human cells in the artificial generation of patients' own tissue, so that new tissue can be grown from them. This tissue could then be transplanted without the risk of rejection. ——— **Intelligent prostheses save lives** A healthy heart is a blessing; a sick heart is life-threatening. If an organ can no longer do its work, a prosthetic organ may be able to take over in the future. Intelligent prostheses fulfil complex tasks in the human organ system. Some ›replacement parts‹ even communicate with the outside world and are, therefore, cleverer than the originals. Which of our organs can already report that it needs to be repaired? ——— **Medicines of tomorrow** Progress in genome sequencing, protein chemistry and bioinformatics help in the innovative development of new medicines. Enormous technical efforts are under way to find new chemical substances for medicines that can fight lethal diseases. Chemists are often already in a position to design suitable active agents and tailor them to specific requirements. ——— **Surgical techniques 2020** With the help of new developments in endoscopy and nanotechnology, methods are being researched that could replace comprehensive surgical interventions. In future, magnetic nanocontainers, controlled by magnetic fields, could deliver an active ingredient directly to the focus of a disease or endoscopic capsules could simply cut diseased cells out of the tissue. ——— **The operating theatre of the future** In the hybrid OP, the surgeon has the support of an innovative X-ray system. Safety for the doctor and patient is improved through constant monitoring. A digital X-ray camera, an OP table that is permeable to X-rays: The doctor can monitor the progress of the operation continuously. The surgery can therefore be carried out with mini-incisions and a heart catheter – for the patient's benefit. ——— **Patient records go digital** With just a few clicks, clinicians can now share key diagnostic and therapeutic patient data across the entire continuum of care. All key data are stored in the electronic patient records and protected by access controls – eHealth solutions prevent duplicate examinations and increase efficiency by improving and accelerating data flows.

1 Der Operationssaal der Zukunft: minimal-invasiver Herzklappenersatz *The operating theatre of the future: endovascular heart valve replacement.* 2 Die künstliche Herzklappe wird per Katheter in die Hauptschlagader eingeführt – dann wird das Drahtgeflecht ausgedehnt, um einen sicheren Halt zu gewährleisten. *The artificial valve is inserted into the aorta using a catheter – then the wire mesh is expanded for a secure hold.* 3 Kunstherz: Durch wechselseitiges Zusammendrücken der beiden Pumpsäcke wird Blut aus den Herzkammern in Lunge und Körperkreislauf gepumpt und das Herz entlastet. *Artificial heart: Through the alternating squeezing of the two pump sacks, the blood is pumped from the heart chambers into lungs and circulatory system and the burden on the heart is alleviated.*

1, 2 Siemens AG   3 DLR

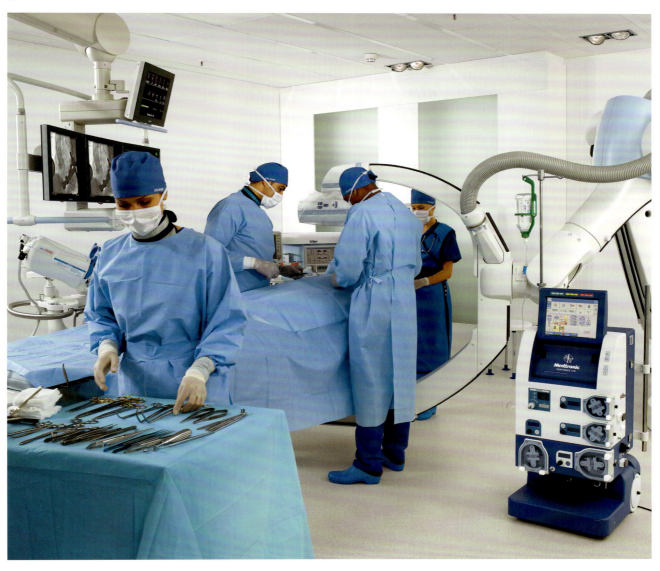

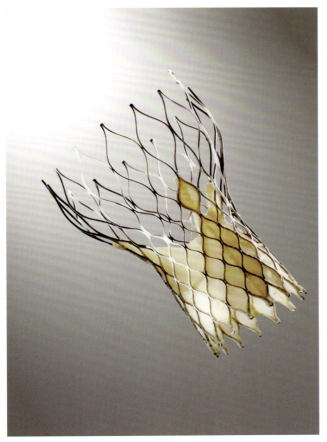

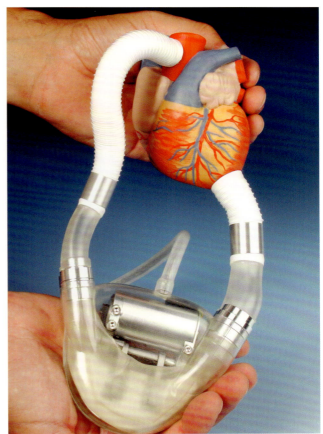

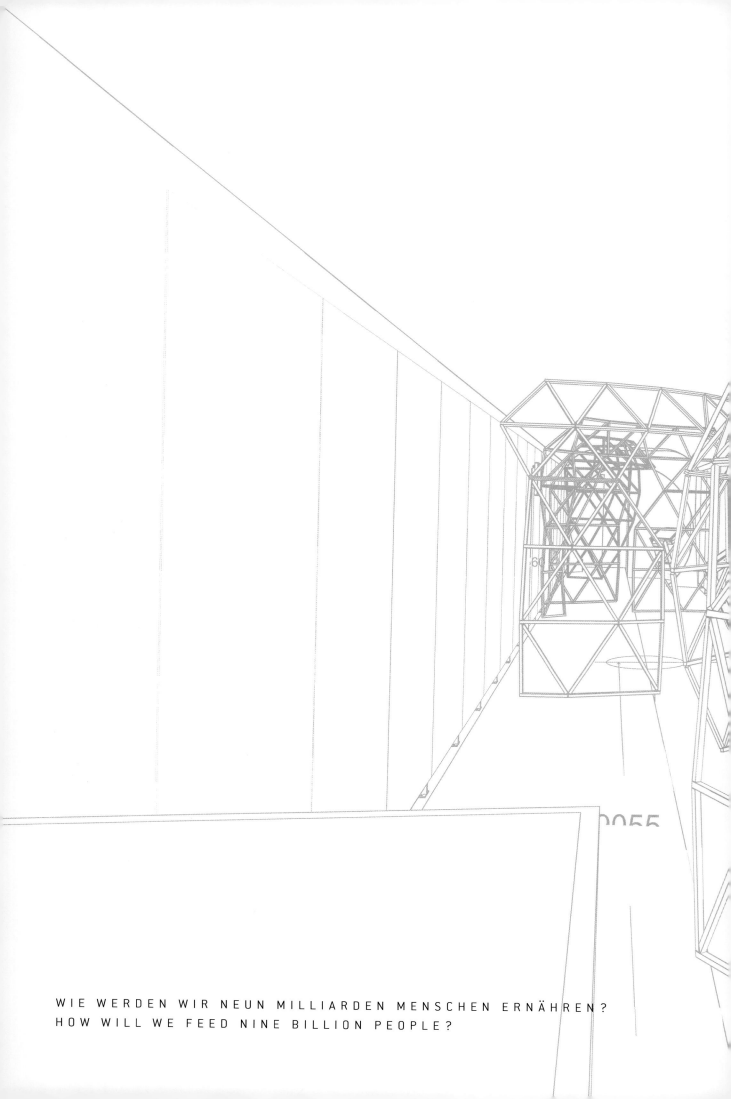

WIE WERDEN WIR NEUN MILLIARDEN MENSCHEN ERNÄHREN?
HOW WILL WE FEED NINE BILLION PEOPLE?

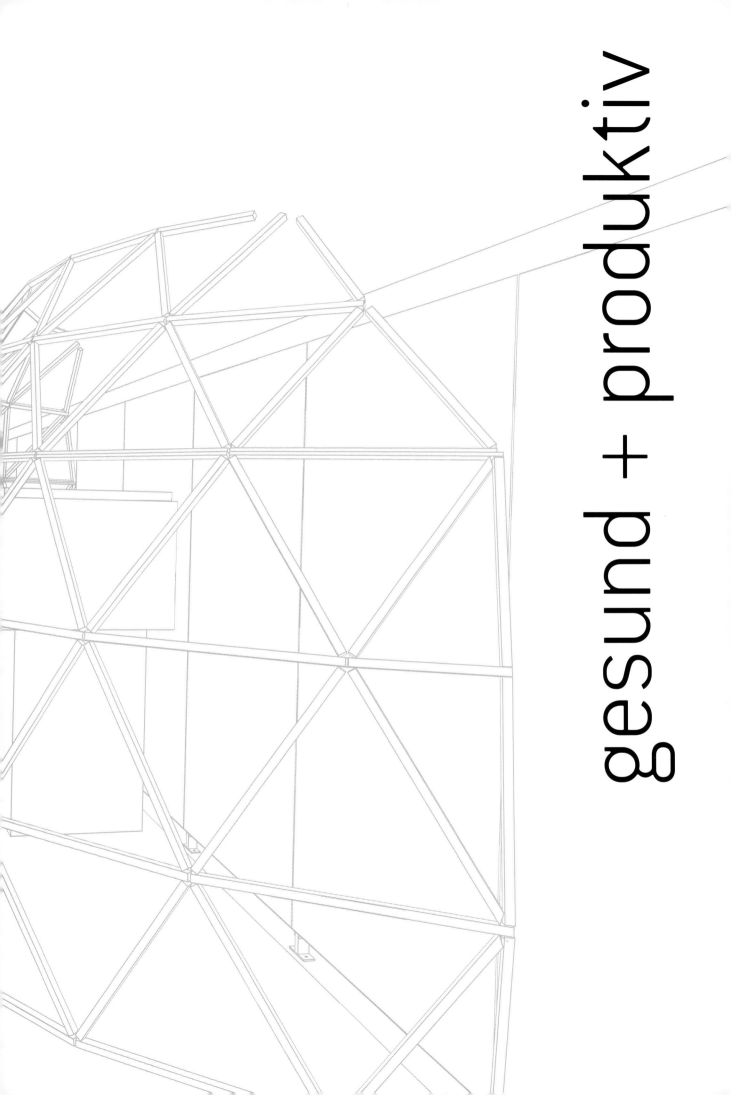

gesund + produktiv

Vorige Doppelseite: Rapspollen auf einem Blütenblatt ─── Hintergrund: Bewässerung in New South Wales: Obwohl hier nur sechs Prozent des Niederschlags in Australien fallen, werden 41 Prozent der Nahrungsmittel produziert. *Background: Irrigation in New South Wales: Although the rainfall is only up to six percent of the whole Australian precipitation, 41 percent of Australia's food is produced here.*
Vorige Doppelseite *Preceding spread* BASF SE Hintergrund *Background* Mit freundlicher Unterstützung von National Geographic

Wie werden wir neun Milliarden Menschen ernähren?

Ackerbau und Viehzucht führten zu einem Wendepunkt in der Menschheitsgeschichte – unsere Vorfahren wurden sesshaft. Seither wächst die Weltbevölkerung stetig und stellt uns heute vor eine gewaltige Herausforderung: noch einige Milliarden mehr Menschen zu ernähren. Wasser, Boden, Klima – unabdingbar sind diese Faktoren mit der Landwirtschaft verknüpft. Der Klimawandel vollzieht sich schneller als erwartet und erschwert es, die Ernährung einer rasant wachsenden Bevölkerung zu sichern. Während große Teile der Menschheit nicht ausreichend versorgt werden, verbreiten sich ernährungsbedingte Krankheiten besonders in den Schwellenländern wie eine Epidemie. Die grüne Biotechnologie und neue industriell erzeugte Lebensmittel versprechen Auswege. Lassen wir uns auf diese Veränderungen ein? ─── **Unser Nahrungsbedarf verändert die Erde** *Die Kluft zwischen Armut und Überfluss, Ressourcenverbrauch und stabilen Ökosystemen wird breiter. Der steigende Nahrungsmittelbedarf verlangt immer größere Ackerflächen und höhere Fangquoten und verursacht irreparable Schäden an der Umwelt. Noch immer ist die Welt nicht frei von Hungersnöten. Gleichzeitig nehmen Schwellenländer schnell westliche Ernährungs- und Verhaltensweisen an und bezahlen den gewonnenen Wohlstand mit mangelnder Gesundheit. Künftig muss es gemeinsame Aufgabe sein, allen Menschen hochwertige Nahrungsmittel zu bieten, ohne der Erde weitere Anbauflächen abzuringen und genutzte Flächen übermäßig zu belasten. Dabei müssen wir vielleicht über uns selbst hinauswachsen.*
─── **Die Welt in Zahlen** *Zahlen lügen nicht. In jeder Sekunde beeinflussen wir die natürlichen Ökosysteme unseres Planeten nachhaltig. Und die Uhr tickt … Die Erdbevölkerung wächst stetig. Ebenso nimmt der Bedarf an Nahrung und Raum beständig zu. Bliebe die weltweite Geburtenrate bis 2050 konstant, würden auf der Erde dann 11,9 Milliarden Menschen leben.* ─── **Wir hinterlassen eindeutige Spuren** *Keine Spezies hat das Angesicht der Erde so nachhaltig verändert wie der Mensch. Unser Handeln wandelt unwiederbringlich Boden, Ozeane und Atmosphäre. Mit Zunahme der Weltbevölkerung schrumpft scheinbar der Planet. Die weißen Flecken werden immer weniger, überall hinterlassen die Menschen ihren Fußabdruck. Wir selbst sind abhängig vom Zustand unserer Erde; ihre Gesundheit liegt in unseren Händen.* ─── **Falscher Lebensstil macht dick** *Deutschland ist dick: 66 Prozent der Männer und 51 Prozent der Frauen sind übergewichtig. Das spiegelt den globalen Trend – mit fatalen Folgen für die Gesundheit. Hunger resultiert aus Armut, denn die Welt produziert eigentlich genug Nahrung. Aber nicht jeder kann sich gesunde Nahrung kaufen. Kehrt mehr Wohlstand ein, greifen die ärmeren Bevölkerungsschichten zu billigen Süßungsmitteln, Fetten und Fleischwaren.*
─── **Perspektiven der Landwirtschaft** *Grüne Hochhäuser, vertikale Farmen und Agroparks: Architekten und Forscher wollen in Zukunft mit komprimierten, autarken Systemen hoch hinaus und an Boden gewinnen. Für eine wachsende Bevölkerung gibt es immer weniger Platz. So entstand die Idee, Farmen in die Höhe zu bauen oder Menschen eines Wohnblocks zu Selbstversorgern zu machen. Agroparks sind durch vernetzte Prozesse energieeffizient und kostengünstig.* ─── **Die Biotechnologie zeigt Wege aus der Krise** *Konventionelle Züchtung schöpft nicht alle Möglichkeiten aus, um unsere Ansprüche an Qualität, Quantität und Umweltschutz zu erfüllen. Die Molekularbiologie gibt den Agrarwissenschaften neue, biotechnologische Werkzeuge, um die Welternährung zu sichern. Die grüne Revolution Anfang der 1960-er Jahre hat die Nahrungsmittelproduktion effizienter gemacht und vermittelt, dass eine ausreichende Versorgung auch für Entwicklungsländer möglich ist. Trotz der Erfolge bleibt die Lage angespannt. Einen Beitrag zur Lösung verspricht auch die Gentechnik. Während genetisch optimierte Kulturen längst angebaut werden, ist diese Technologie gerade in den Industriestaaten in der Kritik. Ihre Anwendung verläuft schneller als ihre ethisch-moralische Integration.* ─── **Züchtung und Biotechnologie sind keine Gegensätze**

Seit 10.000 Jahren erschaffen die Menschen durch Züchtung immer leistungsfähigere Pflanzen, wie sie durch die natürliche Evolution nie entstanden wären. Gezüchtete Kultursorten unterscheiden sich äußerlich und genetisch stark von den ursprünglichen Wildformen. Durch die grüne Gentechnik können neue Eigenschaften schneller und gezielter in das Pflanzengenom eingebracht werden.

Pflanzen werden stressresistenter Pflanzen, die mit weniger Wasser auskommen, unter widrigsten Bedingungen gedeihen oder sogar Wirkstoffe produzieren – die Gentechnik macht es möglich. Der Nahrungsbedarf einer wachsenden Bevölkerung verlangt neue Strategien. Sie betreffen etwa das Welthandelssystem oder die Ernährungsgewohnheiten auch der entwickelten Länder. Ein anderer Ansatz ist es, nahrhaftere Sorten oder solche mit einer Toleranz gegenüber Dürre biotechnologisch zu realisieren. **Der Anbau wird kontrolliert** Noch wissen wir nicht alles über die Wechselwirkung von genetisch veränderten Organismen und ihrer Umwelt. Daher unterliegt der Anbau von Biotech-Pflanzen strengen Kontrollen. Genetisch veränderte Mais- und Sojapflanzen werden weltweit auf riesigen Flächen angebaut, während die Debatten um potenzielle Gefahren, Wagnis und Verantwortung noch nicht abgeschlossen sind. Sind wir reif genug für eine zweite ›grüne Revolution‹? **Was wollen wir essen?** *Genug und gesund: In punkto Welternährung wird zukünftig nicht nur das Wie, sondern auch das Was von großem Interesse sein. Fehlernährung verursacht dem Gesundheitssystem hohe Kosten, denn viele ernste Erkrankungen entstehen durch falsche Ernährung. Die Gesundheit ist unmittelbar mit der Ernährung verknüpft. Unser Körper bildet täglich eine Million neuer Zellen. Die Bausteine dazu liefert unsere Nahrung. Mit diesem Bewusstsein wird der Ruf nach gesunden, auf den persönlichen Bedarf zugeschnittenen Lebensmitteln immer lauter. Das Zeitalter der Genetik macht auch vor der Ernährungsforschung nicht halt und könnte für eine Revolution sorgen. Generieren Automaten unsere – natürlich fettfreien und vitaminhaltigen – Steaks bald aus Einzelmolekülen?* **Auf dem Weg zur individuellen Ernährung** Auf dem Weg zu einer auf den einzelnen Menschen abgestimmten Ernährung gab es viele Stationen. Wo stehen wir heute? Fleischkonsum, Fertigprodukte, Ökowelle: Die Trends der vergangenen Jahrzehnte spiegelten stets die Grundstimmungen der Gesellschaft. In Zeiten der Selbstverwirklichung steht heute die personalisierte, auf den Einzelnen abgestimmte Ernährung hoch im Kurs. **Essen um gesund zu bleiben** Eine individuelle Ernährung resultiert nicht nur aus langjähriger Erfahrung. Auch unsere genetische Ausstattung bestimmt, was für uns gesund ist. Ernährungsratschläge und funktionelle Lebensmittel beruhen oft auf jahrzehntelangen Studien. Doch unsere Nahrung beeinflusst den Körper weitaus mehr, indem sie über Hormone und Stoffwechselprodukte direkt auf die Regulation von Genen einwirkt. **Wir sind, was wir essen** In diesem Spruch steckt mehr Wahrheit als gedacht, werden doch täglich Substanzen in unserer Nahrung entdeckt, die unser Erbgut direkt beeinflussen. Die Nutrigenomik untersucht die Wechselwirkungen zwischen Nahrungsbestandteilen, Essverhalten und dem Wirken unserer Gene. Sie wird uns eines Tages empfehlen, was für den Einzelnen gesund ist und wie wir ernährungsbedingte Krankheiten behandeln können.

1 Kulturen leiden unter abiotischem Stress wie z. B. Dürre. Die Entwicklung von stressresistenteren Pflanzen könnte die Ertragseinbußen minimieren. *Crops suffer from abiotic stress and from drought. The development of stress-resistant plants could minimize yield losses.* 2, 3 Unter UV-Licht offenbaren Tabakpflanzen den Erfolg des Gentransfers; das fluoreszierende Protein dient als Test für spätere Wirkstoffproteine. *Tobacco plants under UV light reveal the success of genetic transfer; the flourescent protein is used to test for desired protein agents.* 4 Pollen (blau) ist der natürliche Überträger von genetischem Material bei Pflanzen. Sie werden in den Staubblättern der Blüte gebildet. *Pollen grains (blue), the natural carriers to exchange genetic information among plants, are produced inside the anthers of the flower.*

1 Bayer AG; Grafik: ArchiMeDes 2, 3 Bayer AG 4 P. Huijser und A. Grande, Max-Planck-Institut für Züchtungsforschung, Köln

1

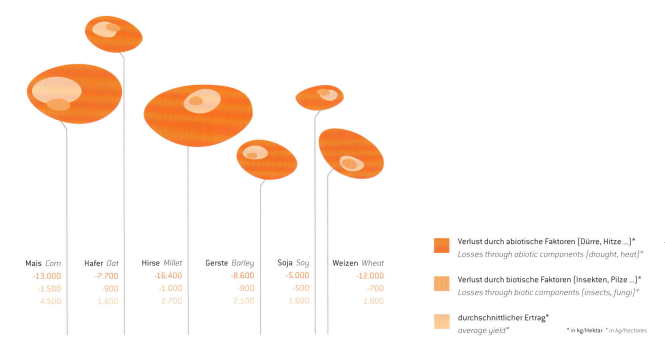

Mais Corn	Hafer Oat	Hirse Millet	Gerste Barley	Soja Soy	Weizen Wheat
-13.000	-7.700	-16.400	-8.600	-5.000	-12.000
-1.500	-900	-1.000	-800	-500	-700
4.500	1.800	2.700	2.100	1.600	1.800

■ Verlust durch abiotische Faktoren (Dürre, Hitze …)*
Losses through abiotic components (drought, heat)

■ Verlust durch biotische Faktoren (Insekten, Pilze …)*
Losses through biotic components (insects, fungi)

■ durchschnittlicher Ertrag*
average yield   * in kg/Hektar   * in kg/hectares

2

3

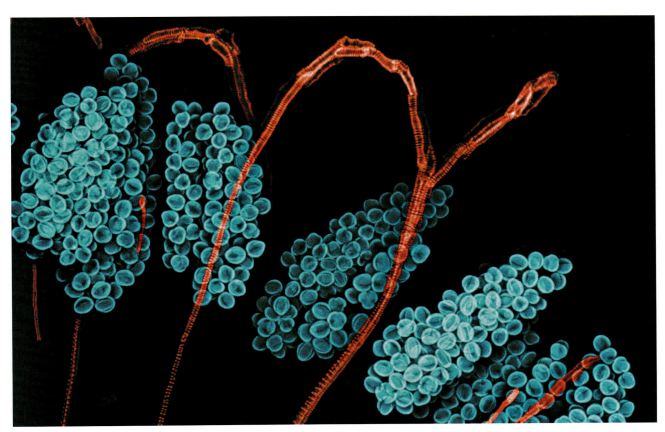

4

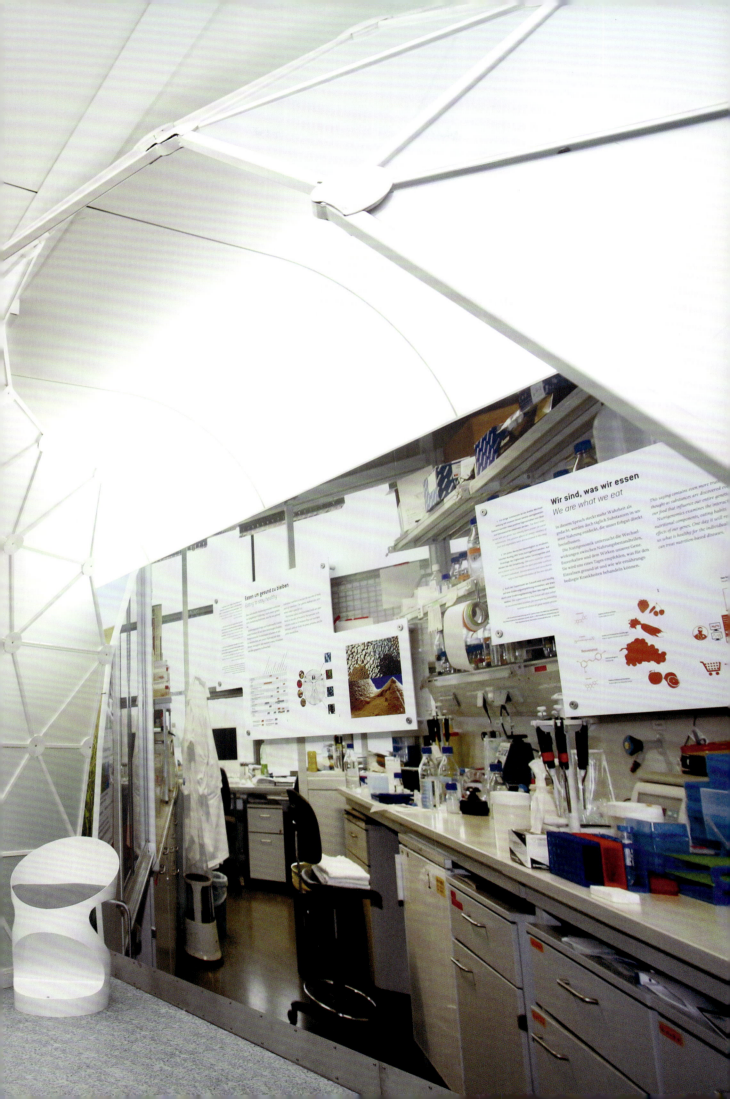

Übersicht der Verbreitung und heutigen Fundorte des Weizens
Map of current natural distribution and today's location of wheat

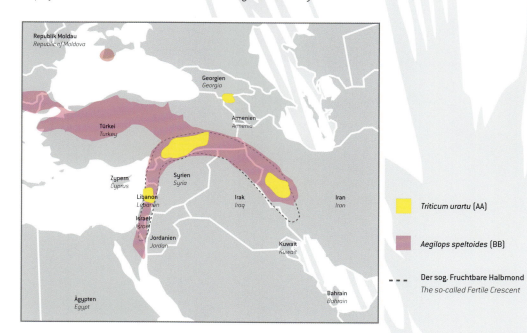

- Triticum urartu (AA)
- Aegilops speltoides (BB)
- --- Der sog. Fruchtbare Halbmond / *The so-called Fertile Crescent*

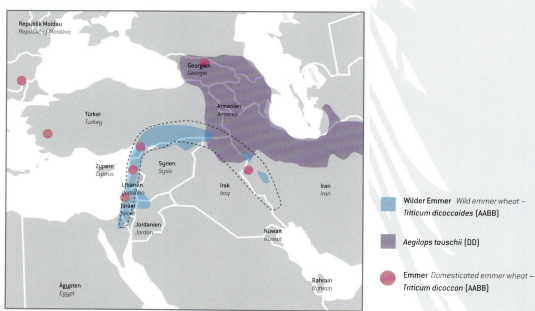

- Wilder Emmer *Wild emmer wheat –* Triticum dicoccoides (AABB)
- Aegilops tauschii (DD)
- Emmer *Domesticated emmer wheat –* Triticum dicoccon (AABB)

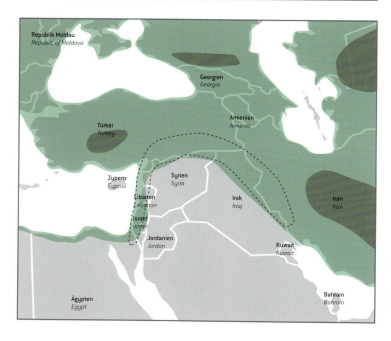

- Hartweizen *Durum wheat –* Triticum durum (AABB)
- Brotweizen *Bread wheat –* Triticum aestivum (AABBDD)
- Keine Weizenanbaugebiete / *No wheat cultivation*

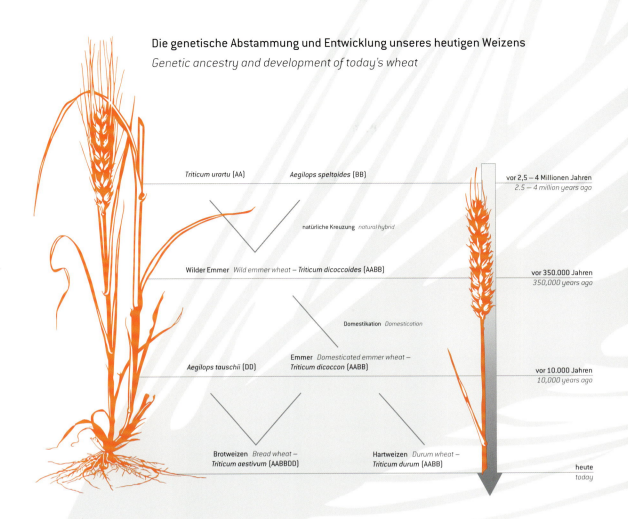

Die genetische Abstammung und Entwicklung unseres heutigen Weizens
Genetic ancestry and development of today's wheat

Triticum urartu (AA)   *Aegilops speltoides* (BB)   vor 2,5 – 4 Millionen Jahren / *2.5 – 4 million years ago*

natürliche Kreuzung *natural hybrid*

Wilder Emmer *Wild emmer wheat* – *Triticum dicoccoides* (AABB)   vor 350.000 Jahren / *350,000 years ago*

Domestikation *Domestication*

Aegilops tauschii (DD)   Emmer *Domesticated emmer wheat* – *Triticum dicoccon* (AABB)   vor 10.000 Jahren / *10,000 years ago*

Brotweizen *Bread wheat* – *Triticum aestivum* (AABBDD)   Hartweizen *Durum wheat* – *Triticum durum* (AABB)   heute / *today*

How will we feed nine billion people?

Tillage and livestock breeding lead to a turning point in the history of humanity – our predecessors became sedentary. Since then, the world's population has been growing constantly and poses an enormous challenge today: how do we feed another few billion people? Water, soil, climate – three factors that are inextricably linked with agriculture. Climate change is happening faster than expected and this makes it difficult to secure the food supply of a rapidly increasing global population. While large sectors of humanity are inadequately supplied with food, nutrition-based illnesses are spreading like an epidemic, particularly in threshold countries. Green biotechnology and new industrially produced foods offer possible solutions. The question is: will we allow these changes to happen? **Our food requirement is changing the Earth** *The gap between poverty and excess, the consumption of resources and stable ecosystems is growing. The increasing demand for food requires ever bigger areas of arable land and higher fishing quotas, and is causing irreparable damage to the environment. Our world is still facing famine. At the same time, threshold countries are rapidly adopting Western nutritional and behavioural habits and are paying for the gains in prosperity with health problems. In future we must share the task of providing all people with high-quality food without creating new arable areas, and excessively exploiting the areas already in use. To do this we must, perhaps, grow beyond ourselves.* **The world in numbers** Numbers do not lie. Every second we make a lasting impact on our planet's ecosystems. And the clock is ticking … The world's population is growing incessantly. The need for food and space is also increasing. If the global birth rate were to remain constant until 2050, there would be 11.9 billion people on Earth. **We leave clear and lasting traces behind** No species has changed the face of the Earth as profoundly as man. Our actions cause irreversible changes to the land, oceans and atmosphere. Due to the increase in the world's population, the planet appears to shrink. There are fewer

1

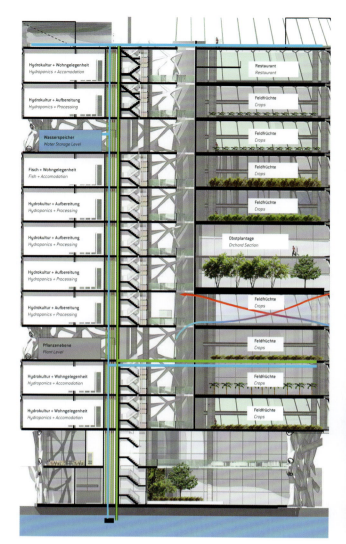

2
3

Das Gesundheits-Servicepaket *Health Service Plan*

4

1 USA 62,5 Millionen Hektar *United States 62.5 million hectares*	**3** Brasilien 15,8 Millionen Hektar *Brazil 15.8 million hectares*	**5** Kanada 7,6 Millionen Hektar *Canada 7.6 million hectares*	**7** Paraguay 2,7 Millionen Hektar *Paraguay 2.7 million hectares*
2 Argentinien 21 Millionen Hektar *Argentina 21 million hectares*	**4** Indien 7,6 Millionen Hektar *India 7.6 million hectares*	**6** China 3,8 Millionen Hektar *China 3.8 million hectares*	**8** Südafrika 1,8 Millionen Hektar *South Africa 1.8 million hectares*

■ Kommerzieller Anbau von gentechnisch verändertem Getreide
Grows GM crops commercially

■ Verbot kommerziellen Anbaus von gentechnisch verändertem Getreide
Prohibits commercial growth of GM crops

■ Einfuhr von gentechnisch verändertem Getreide als Nahrung oder Futter erlaubt
Allows import of GM crops for food and/or feed

and fewer blank spots on the map, humans are leaving their footprints everywhere. We are dependent on the condition of our Earth; its health is in our hands. **A poor lifestyle is fattening** Germany is fat: 66 percent of men and 51 percent of women are overweight. This reflects the global trend – with fatal consequences for health. Hunger is not the result of poverty. The world actually produces enough food. However, not everyone is able to buy healthy food. When greater affluence arrives, the poorer sectors of the population resort to cheap sweeteners, fats and meat products. **Prospects for agriculture** Green high rises, vertical farms and agroparks: Architects and researchers are reaching to the sky with high-density, self-sufficient systems and, in this way, gaining space. There is increasingly less space available for the world's growing population. Consequently, the idea has emerged of building farms vertically and helping the people in a residential block to become self-sufficient. Thanks to networked processes, agroparks are energy efficient and cost-efficient. **Biotechnology reveals ways out of the crisis** *Conventional animal breeding will not suffice on its own to fulfil future requirements in relation to quality, quantity and environmental protection. Molecular biology provides new biotechnological tools to help secure global food supply. The Green Revolution in the early 1960s made food production more efficient and conveyed the message that a sufficient food supply is possible, even in developing countries. Despite associated successes, however, the situation remains difficult. While genetic technology provides some solutions and genetically modified crop has been grown for some time now, this technology is subject to criticism, particularly in industrialised countries. Its practical implementation is running more smoothly than its ethical acceptance.* **Breeding and biotechnology are not opposites** People have been creating more effective plants through breeding for 10,000 years in a way that could never have arisen solely through natural evolution. Cultivated plant varieties differ significantly from the original wild forms,

1 Vertical Farming hat das Potenzial, Abholzung, Urbanisierung und Umweltverschmutzung durch Transporte, Chemikalien und andere Pestizide zu reduzieren. Diese Vertical Farm könnte 6.000 Menschen versorgen, die sich vegetarisch und von Fisch ernähren. *Vertical Farming holds the potential to reduce deforestation, urban sprawl and pollution by dramatically reducing transport, chemicals and other pesticides. This proposed Vertical Farm aims to serve 6,000 on a vegetarian diet and fish.* 2 Übergewicht hat viele Gründe: In Mäusen, die unter Fettsucht leiden, wurden bereits Gene lokalisiert, die mit dem Krankheitsbild assoziiert sind. *Fatness has many causes: genes associated with this condition have already been identified in mice suffering from obesity.* 3 Die Trennung pflanzlicher Rohstoffe in Öl, Proteine, Kohlenhydrate und sekundäre Pflanzenstoffe ermöglicht den Ersatz ungesunder Inhaltsstoffe. *The separation of plant raw materials into oils, proteins, carbohydrates and secondary plant materials enables us to replace unhealthy substances.* 4 Um jeden Menschen bestmöglich zu ernähren, könnten künftig DNS-Tests durchgeführt werden. Auf Grundlage der Ergebnisse helfen Servicegeräte und Berater bei der Umsetzung. *DNA tests could be carried out in future so as to provide the best possible nutrition for each individual. Service devices and consultants help in the implementation of the test results.* 5 Auf der ganzen Welt wird bereits genetisch verändertes Getreide angebaut. Spitzenreiter sind die USA, gefolgt von Ländern Südamerikas und Kanada. *Genetically modified grain is already grown throughout the world. The USA are the pioneer in this sector, followed by several South American countries and Canada.*

1 Oliver Foster, 0 Design, www.0design.com.au 2 Deutsches Institut für Ernährungsforschung, Potsdam-Rehbrücke (DIfE) 3 Fraunhofer-Institut für Verfahrenstechnik und Verpackung IVV, Freising 4 Mit freundlicher Unterstützung von Prof. Hannelore Daniel, Technische Universität München

both physically and genetically. Green genetic technology allows for faster and targeted introduction of desired characteristics into the plant genome. **More crop per drop** Plants that can survive on less water thrive under the most hostile conditions and even produce active substances – genetic technology makes it all possible. New strategies are required to meet the food requirement of our growing population. They concern the world trades system or the food patterns also of the developed countries. Alternatively more nutritious, drought tolerant varieties can be accomplished with the help of biotechnology. **Cultivation is controlled** We still do not fully understand the interactions between genetic modified organisms and their environment. For this reason, the cultivation of biotech plants is subject to very strict controls. Genetically-modified corn and soy plants are grown throughout the world on a vast scale, while the debates surrounding the potential hazards, risks and responsibility are still under way. Are we mature enough for a second ›green revolution‹? **What do we want to eat?** *Enough and healthily: When it comes to feeding the world, in future there will be enormous interest not just in the ›how‹ but also the ›what‹. Poor nutrition generates high costs for the healthcare system; many serious illnesses result from bad nutrition. Health is directly linked with nutrition. Our bodies generate a million new cells every day. Our nutrition provides the building blocks for this process. Based on this, the call for healthy food tailored to personal requirements is becoming more urgent. The age of genetics does not stop at nutrition research and could provoke a revolution. Will our steaks – fat-free and full of vitamins, of course – soon be generated from individual molecules by machines?* **On the path to personalised nutrition** There have been many stops along the way to nutrition tailored by the individual. So where do we stand today? Meat consumption, processed foods, the eco-wave: The trends of recent decades continually reflected the basic moods of society. Personalised nutrition, which is tailored to the needs of the individual, is highly rated in these times of self-realisation. **Eating to stay healthy** Individual nutrition is not just the outcome of many years of experience. Our genetic makeup also determines what is good for us. Nutritional advice and functional foods are often based on decades of research. However, our nutrition influences the body far more than this as it has a direct effect on the regulation of our genes through hormones and metabolic products. **We are what we eat** This saying contains even more truth than we thought as substances are discovered every day in our food that influence our entire genetic makeup. Nutrigenomics examines the interaction between nutritional components, eating habits and the effects of our genes. One day it will recommend to us what is healthy for the individual and how we can treat nutrition-based diseases.

IN KREISLÄUFEN DENKEN — RESSOURCEN SCHONEN
THINKING IN CYCLES — SAVING RESOURCES

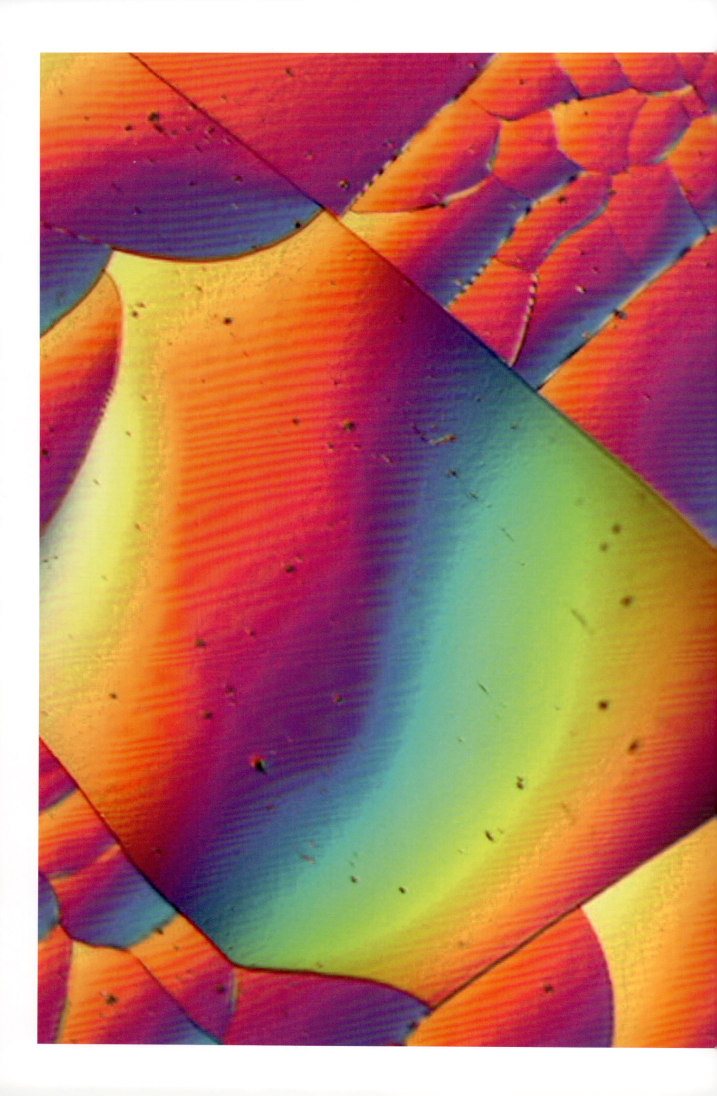

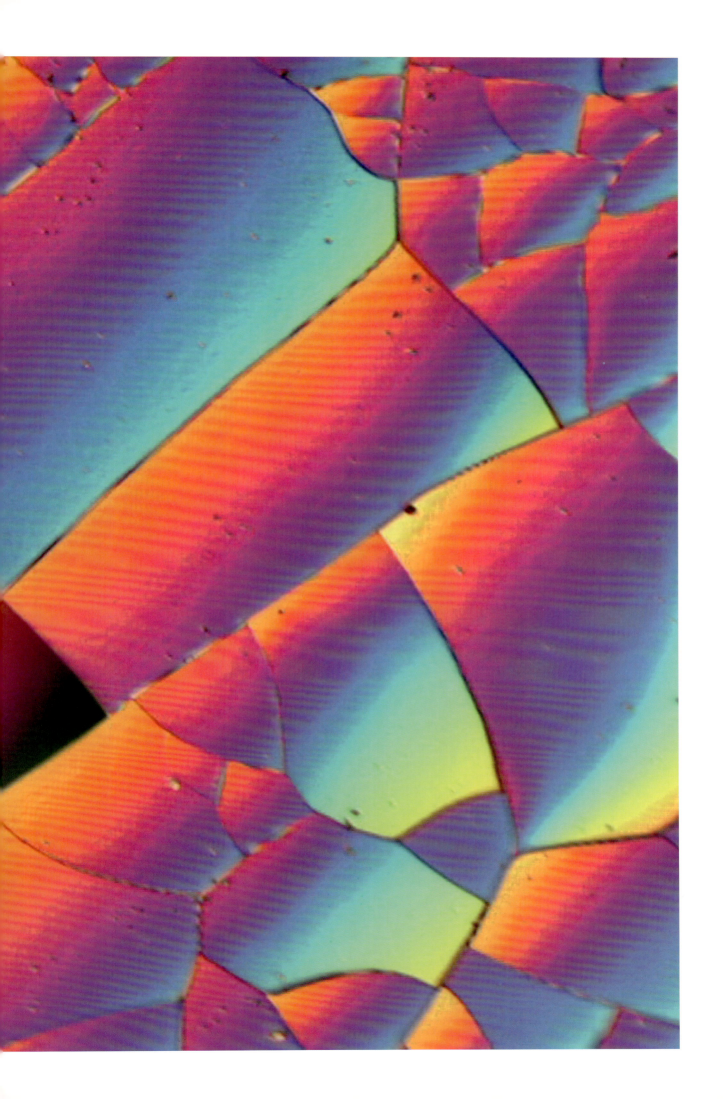

1
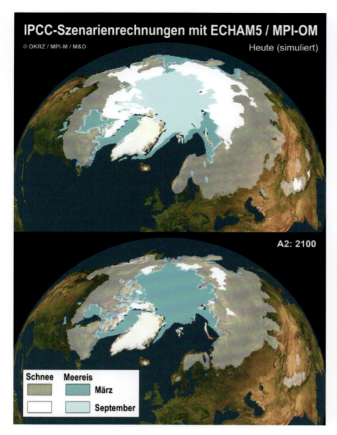

2

3

In Kreisläufen denken – Ressourcen schonen

Durch immer genauere Beobachtungen verstehen wir unsere Welt immer besser. Wir wissen, dass der Mensch das Klima beeinflusst und dass die Energieversorgung keineswegs gesichert ist. Wie werden wir in Zukunft mit der Erde und ihren Reserven umgehen? Fast unsere gesamte Energie beziehen wir von der Sonne, entweder direkt oder über den Umweg von Wind, Wasser und – fossiler oder nachwachsender – Biomasse. Einzig die Nutzung von Gezeiten und Erdwärme gehen auf den Mond bzw. die Erde selbst zurück. Wir müssen lernen, Energie effizienter zu nutzen, sie nur dort zu verbrauchen, wo es wirklich nötig ist, und in Kreisläufen zu denken, um das labile Gleichgewicht unseres Planeten nicht noch mehr zu stören.

Das Klima der Erde ändert sich *Klimaprognosen sind heute durch den Vergleich von Modellrechnungen mit Beobachtungsdaten verlässlich und detailliert. Der Einfluss des Menschen auf das Klima ist unbestritten und neue Modelle zu sozioökonomischen Auswirkungen werden entwickelt. Heutige Modelle für das Klima in 50 oder 100 Jahren umfassen Atmosphäre, Ozean, Kryo- und Biosphäre und können durch Vergleiche mit immer genaueren, aktuellen Beobachtungsdaten und historischen Klimaaufzeichnungen überprüft werden. Kurzfristige und detaillierte, regionale Prognosen sind derzeit noch durch Rechenkapazitäten und unverstandene Rückkopplungsprozesse begrenzt. Neben der globalen Temperaturerhöhung gewinnen Vorhersagen zur regionalen Klimaentwicklung immer mehr an Bedeutung.*

Die Erde – Schicht für Schicht Neue Methoden und Instrumente erlauben es, das Erdsystem heute aus den verschiedensten Blickwinkeln zu betrachten – viel detaillierter als je zuvor. Sonden und Satelliten vermessen die Erde bis zwölf Kilometer Tiefe und 1.000 Kilometer Höhe. Dabei liefern sie detaillierte und zum Teil zeitaufgelöste Daten über viele Bestandteile der Atmosphäre sowie über Strömung, Temperatur und Zusammensetzung der Ozeane.

Traditionelle Energieerzeugung wird sauberer *Auch wenn alternative Formen der Energieerzeugung heute oft im Fokus der Energiediskussion stehen, werden wir noch auf absehbare Zeit auf die effiziente Ausbeute herkömmlicher Energiequellen angewiesen sein. Die Zukunft von Kohle, Gas und Öl hängt davon ab, ob es gelingt, das klimaschädliche Kohlendioxid mit neuen Methoden aus der Atmosphäre fernzuhalten oder es zu neutralisieren. Außerdem werden innovative Verfahren entwickelt, um neue oder bisher unwirtschaftliche Lagerstätten zu erschließen. Für Kernkraftwerke liegt das Problem nicht im emittierten Kohlendioxid, sondern im radioaktiven Abfall. Dies könnten Reaktoren der vierten Generation deutlich mindern, da sie strahlenden Abfall mit erheblich kürzeren Halbwertszeiten produzieren.*

Energie aus Kohle & Co Mehrere technische Verfahren, um beim Verbrennen von fossilen Energieträgern entstehendes Kohlendioxid abzuspalten, befinden

---

Vorige Doppelseite: Material für zukünftige Fusionsanlagen: Untersuchung einer Wolframsilizid-Schicht auf Quarzglas. 1 Bei weiter steigendem Kohlendioxid-Ausstoß ist davon auszugehen, dass das Meereis der Nordpolregion im Sommer vollständig abschmilzt. 2 Mit dem 300 Meter hohen Messturm ZOTTO in der sibirischen Taiga können Treibhausgase sowohl lokal als auch über weiträumigen Gebieten gemessen werden. 3 Die vier Meteosat-Satelliten befinden sich auf geostationären Umlaufbahnen und beobachten das Wetter mit hoher Abtastfrequenz in zwölf Spektralbändern. 4 Das wichtigste Werkzeug der deutschen Polarforschung: Das Forschungs- und Versorgungsschiff Polarstern.

Preceding spread: Material for future fusion plants: study of a tungsten silicide layer on quartz glass. 1 If carbon dioxide emissions continue to rise, model calculations indicate that the sea ice in the Arctic region will melt completely during the summer. 2 The 300 metre ZOTTO tower in the Siberian taiga can measure greenhouse gases locally and over large areas. 3 The four Meteosat satellites are in geostationary orbits and monitor weather patterns with frequent sampling, every 15 minutes, in twelve spectral bands. 4 The most important tool for German polar research: The Polarstern research and supply vessel.

Vorige Doppelseite Gabriele Matern, Max-Planck-Institut für Plasmaphysik, Garching 1 DKRZ, Max-Planck-Institut für Meteorologie, Hamburg, M & D 2 Max-Planck-Institut für Biogeochemie, Jena 3 EUMETSAT 4 Alfred-Wegener-Institut für Polar- und Meeresforschung, Bremerhaven; Foto: Ingo Arndt

sich im Versuchsstadium. Das produzierte CO_2 muss aber nicht unbedingt vom Verursacher selbst neutralisiert werden – durch den Handel mit Emissionszertifikaten können Aufforstung oder andere Programme zur CO_2-Reduktion finanziert werden. **Hilft uns brennendes Eis?** Am Meeresboden ruht ein gigantisches Energiepotenzial: Methanhydrat, eine eisähnliche Verbindung aus Wasser und Methan, dem Hauptbestandteil von Erdgas. Schätzungen ergeben, dass in ›Methaneis‹ doppelt so viel Kohlenstoff zu finden ist, wie in allen bekannten Erdgas-, Öl- und Kohlevorkommen zusammen. Der Abbau des Hydrats ist schwierig, da es nur bei hohem Druck und niedrigen Temperaturen stabil ist – entweicht es, fördert es den Treibhauseffekt massiv. **Kernkraftwerke der vierten Generation** Ein von einem Neutronenbeschleuniger gesteuerter Kernreaktor könnte den radioaktiven Abfall von herkömmlichen Kernkraftwerken deutlich verändern. Durch den Beschuss mit Neutronen werden langlebige Nuklide in geeigneten Anlagen in kurzlebigere oder stabile Nuklide umgewandelt. Die Lagerzeit für radioaktive Abfälle würde sich dadurch von Millionen auf einige Hundert Jahre verringern. **Neue Energiequellen werden effizienter** *Solarzellen werden kleiner und effizienter, Windkraftanlagen größer und leistungsstärker, Energie aus Biomasse umweltfreundlicher. In Deutschland stieg der Anteil der erneuerbaren Energien am Stromverbrauch zwischen 1997 und 2007 von 4,5 auf 14,2 Prozent. Dieser Anteil wird noch weiter wachsen. Erneuerbare Energien sind im Stromnetz allerdings nicht unproblematisch: Die Energieerzeugung hängt von der Tageszeit oder den Wetterverhältnissen ab und lässt sich nur schwer für den späteren Verbrauch speichern. Künftig werden intelligente Netze, in denen dezentrale Kapazitäten zu virtuellen Kraftwerken verschaltet sind, helfen, diese Probleme zu lösen.* **Die Sonne im Reaktor** Die Kernfusion könnte in einigen Jahrzehnten als umweltfreundliche und sichere Energiequelle den Energiemix bereichern. Das Fusionsexperiment ITER soll zehnmal so viel Energie erzeugen, wie man zum Aufheizen des Brennstoffs aufwenden muss. Um das Plasma auf den extrem Temperaturen höher als im Innern der Sonne zu halten, sind speziell geformte Magnetfelder nötig. **Strom und Wärme von der Sonne** In thermischen und photovoltaischen Anlagen wird Sonnenenergie bereits vielfach direkt genutzt, sowohl für die Warmwasser- als auch zur Stromerzeugung. Neue Materialien und technische Entwicklungen erlauben einen immer einfacheren Einsatz von Solaranlagen, und das mit steigender Effizienz. Großflächige solarthermische Anlagen zur Stromerzeugung sind in Planung oder bereits im Bau. **Was ist mit Biostrom und Biosprit?** Landwirtschaftliche Flächen müssen für Nahrungsmittel reserviert bleiben – Bioenergie kann man heute aus biologischen Abfallstoffen erzeugen. Energiereiche Pflanzen für Bio-Ethanol haben durch den dafür nötigen Dünger einen negativen Einfluss auf das Klima. Verfahren der zweiten und dritten Generation können Bioenergie aus Abfallholz, Stroh und sogar aus biologischen Küchenabfällen gewinnen. **Energie aus Wasserstoff?** Im Prinzip speicherbar und sauber beim Verbrennen: Wasserstoff wäre der ideale Energieträger. Effiziente Brennstoffzellen gibt es bereits. Doch noch ist die Speicherung ein Problem: Kompression, Verflüssigung und Transport sind extrem energie- und materialaufwändig. Deshalb wird verstärkt nach chemischen Speichern mit flüssigen oder festen Trägerstoffen gesucht. **Energie effizient nutzen** Am billigsten und umweltfreundlichsten ist die Energie, die wir gar nicht erst verbrauchen. *Carlo Rubbia, Physik-Nobelpreisträger 1984* Energie kann in vielen Bereichen eingespart werden: nicht nur durch effizientere Maschinen und Geräte beim Endverbraucher, sei es im Haushalt oder in der Industrie, sondern auch bei der Energieerzeugung und beim Transport. **Der Mensch verändert das ökologische Gleichgewicht** *Sauberes Wasser, reine Luft und intakte Lebensräume sollten Ziele einer nachhaltigen Nutzung unserer Erde sein. Den Status quo*

---

1 Neuartige Probennehmer, die druckbeständig sind, werden in der Forschung eingesetzt, um auch die Deponierung von CO_2 in Methanhydratvorkommen am Meeresboden zu untersuchen. *Innovative sampling devices, used in research to study the disposal of CO_2 in methane hydrate deposits on the sea floor.* 2 Brennendes Eis: Nach ihrer Bergung vom Meeresboden werden die Gashydrate instabil und zerfallen in Wasser und Methan. Das frei werdende Methan verbrennt mit konstanter Flamme, wenn es entzündet wird. *Burning ice: Following extraction from the seabed, the gas hydrates become unstable and break down into water and methane. The released methane burns with a constant flame when ignited.* 3 Durch den Beschuss von Flüssigmetall mit Protonen werden im MEGAPIE-Experiment Neutronen erzeugt, die sich dazu eignen, radioaktive Abfälle zu ›verbrennen‹. Im Flüssigmetall-Labor KALLA wurden strömungstechnische Versuche für das MEGAPIE-Experiment durchgeführt. *In the MEGAPIE experiment, bombarding liquid metal with protons creates neutrons that can be used to ›incinerate‹ radio-active waste. Flux tests for the MEGAPIE experiment were carried out at the KALLA liquid metal laboratory.* 4 Neue Technologien erschließen auch unrentable Erdölfelder. Hier wird Ölsand durch elektrische Induktion erhitzt und daraus – Energie und Wasser sparend – Bitumen gewonnen. *New technologies enable even unprofitable oil fields to be mined. Here, oil sand is heated by induction – saving energy and water – allowing bitumen to be extracted.*

1 Leihgeber *On loan by* Universität Bremen 2 MARUM, Universität Bremen 3 Karlsruher Institut für Technologie KIT 4 Siemens AG

1

2

3

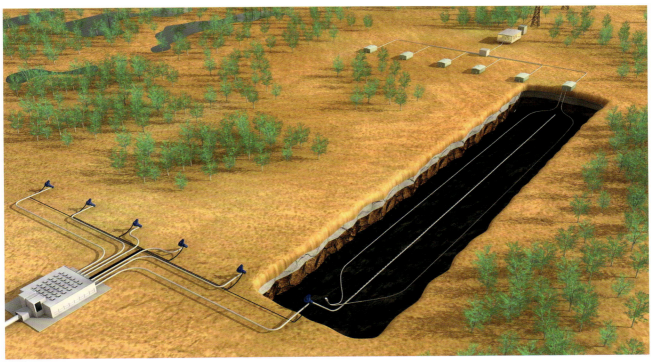

4

Primärenergieverbrauch in Deutschland 2007 *Primary energy consumption in Germany 2007*

Arbeitsgemeinschaft Energiebilanzen, Bundesministerium für Wirtschaft und Technologie, Stand: August 2008

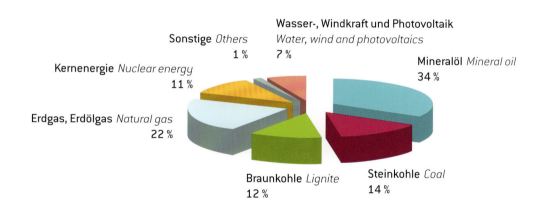

- Sonstige *Others* 1 %
- Wasser-, Windkraft und Photovoltaik *Water, wind and photovoltaics* 7 %
- Kernenergie *Nuclear energy* 11 %
- Mineralöl *Mineral oil* 34 %
- Erdgas, Erdölgas *Natural gas* 22 %
- Braunkohle *Lignite* 12 %
- Steinkohle *Coal* 14 %

Aufteilung des Beitrags erneuerbarer Energien 2007 *Breakdown of renewable energy contributions 2007*

Arbeitsgemeinschaft Energiebilanzen, Bundesministerium für Wirtschaft und Technologie, Stand: August 2008

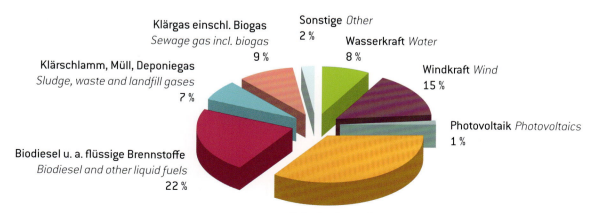

- Klärgas einschl. Biogas *Sewage gas incl. biogas* 9 %
- Sonstige *Other* 2 %
- Wasserkraft *Water* 8 %
- Klärschlamm, Müll, Deponiegas *Sludge, waste and landfill gases* 7 %
- Windkraft *Wind* 15 %
- Photovoltaik *Photovoltaics* 1 %
- Biodiesel u. a. flüssige Brennstoffe *Biodiesel and other liquid fuels* 22 %
- Holz, Stroh u. a. feste Stoffe *Wood, straw and other solid matter* 36 %

1 Mit Wärmespeichern lässt sich die Stromproduktion von der Sonneneinstrahlung entkoppeln. Diese solarthermische Anlage in Almería, Spanien, liefert 230 Kilowatt. *Thermal storage systems allow electricity to be produced whether the sun is shining or not. This solar thermal plant in Almería, Spain, produces 230 kilowatts.* — 2 Im Fusionstestreaktor ITER wird das Plasma durch ein ringförmiges Magnetfeld eingeschlossen. Bei Temperaturen über 100 Millionen Grad produziert es 500 Megawatt. *In the ITER fusion test device, the plasma is confined in a ring-shaped magnetic field. At temperatures in excess of 100 million degrees it will produce 500 megawatts.*

1 DLR/Steur 2 Leihgeber *On loan by* Max-Planck-Institut für Plasmaphysik, Garching

zu erhalten, ist nicht mehr genug, um die Lebensgrundlage der Menschen auch in Zukunft zu sichern. Vielversprechend sind kleinräumige Ansätze des Umweltschutzes: Stärker als bisher werden auf lokaler Ebene Umweltverschmutzung und die Auswirkungen des Klimawandels bekämpft. Globale Maßnahmen zur Kompensation des Treibhauseffekts, wie das ›Geo-Engineering‹ sind schwer realisierbar und meist mit gravierenden Nebenwirkungen verbunden. Eine Abdunklung der Erde durch Schattenspender im All oder die gezielte Düngung des Meeres zur Speicherung von CO_2 können unabsehbare Folgen für das Weltklima haben. **Sauberes Wasser ist Leben** Mehr als eine Milliarde Menschen auf der Erde haben keinen Zugang zu sauberem Trinkwasser. Kleine Projekte haben oft große Wirkung. Dezentrale und technisch einfache Lösungen zur Abwasserbehandlung können nicht nur sauberes Trinkwasser erzeugen, sondern oft auch wertvollen Dünger. So sind viele Abwässer mit Phosphat belastet – ein Rohstoff, dessen Lagerstätten in wenigen Jahrzehnten erschöpft sein werden. **Kleine Poren für reine Luft** Stickstoffdioxid, Feinstaub und Ozon sind die Hauptursachen für Luftverschmutzung, unter der noch immer viele Länder leiden. Abgase stammen hauptsächlich von Kraftwerken, der Schwerindustrie und vom Verkehr. Bessere Filter und modernere Industrieanlagen können dafür sorgen, dass schädliche Gase nicht in die Umwelt gelangen oder in relativ harmlose Stoffe umgewandelt werden. **Lebensräume schaffen und erhalten** Biodiversität ist nicht nur im Regenwald wichtig; auch vor unserer Haustüre gibt es schützenswerte Ökosysteme. In Europa hat der Mensch die Landschaft schon vor Jahrhunderten so verändert, dass es praktisch keine ursprünglichen Lebensräume mehr gibt. In unserer Kulturlandschaft haben sich neue Ökosysteme gebildet, die wieder eine erstaunliche Artenvielfalt zeigen.

Thinking in cycles – saving resources

With increasingly precise scientific observations we now understand our world better than ever. We know that human activity influences the climate and that our future energy supplies are by no means secure. How will we manage the Earth and its resources in the future? We draw virtually all of our energy from the sun, either directly or by way of wind, water and – fossil or renewable – biomass. Only when we use the tides or geothermal heat does the energy originate from the moon or the Earth itself. We must learn to use energy more efficiently, to use it only where we really need to, and to think in cycles so that we can stop disrupting our planet's delicate balance.

The Earth's climate is changing *Climate forecasts are reliable and detailed in this day and age because the model calculations can be compared against observation data. Humankind's influence on the climate is indisputable and new models of socio-economic impact are being developed. Present-day models of the climate in 50 or 100 years' time encompass the atmosphere, oceans, cryosphere and biosphere. These can be verified by comparing them against ever more precise, current observation data and historical climate records. Detailed short-range regional forecasts are currently restricted by limited computing capacities and by feedback processes that we do not yet understand. Besides the rise in global temperatures, forecasts of regional climate trends are becoming increasingly important.*

The Earth – layer by layer New methods and tools now enable us to view the Earth system from almost any imaginable perspective – in much more detail than ever before. Probes and satellites survey the Earth down to a depth of twelve kilometres and up to an altitude of 1,000 kilometres. They deliver detailed and even sometimes time-resolved data on many elements of the atmosphere and on the currents, temperature and composition of the oceans.

Traditional energy generation becomes cleaner Although the energy discussion often focuses on alternative ways of generating energy, we will need to rely on the efficient recovery of energy from conventional sources for the foreseeable future. The future of coal, gas and oil depends upon whether or not we succeed in finding new methods of keeping damaging carbon dioxide out of the atmosphere or neutralising it. Efforts are also being made to develop innovative ways of tapping new or presently uneconomical energy stores. The problem for nuclear power plants is not carbon dioxide but radioactive waste. Fourth generation reactors could provide some relief here, as the radio-active waste they produce has a much shorter half-life.

Energy from fossil fuels Numerous techniques for separating off the carbon dioxide produced during the combustion of fossil-based energy sources are currently being tested. The resulting CO_2 need not necessarily be neutralised by the polluter himself. Countries can trade their emissions certificates, helping to finance reforestation or other CO_2 reduction programmes.

Could burning ice help us? There are enormous reserves of stored energy resting on the seabed in the form of methane hydrate, a solid compound of water and methane, the main component of natural gas. According to estimates, ›methane ice‹ could contain double the amount of carbon than all known reserves of natural gas, oil and coal taken together. But mining the hydrate is difficult, as it is only stable under high pressure and at low temperatures. Escaping methane contributes strongly to the greenhouse effect.

Fourth generation nuclear power plants A nuclear reactor controlled by a neutron accelerator could alter the radioactive waste produced by conventional nuclear power plants. Long-lived nuclides can be transformed into shorter-lived or stable nuclides by bombarding them with neutrons inside suitable plants. This would cut the storage time for radioactive waste from millions of years to several hundred.

New energy sources become more efficient *Solar cells are becoming smaller and more efficient, wind turbines are becoming larger and more powerful, energy from biomass is increasingly environmentally friendly. The contribution of renewable*

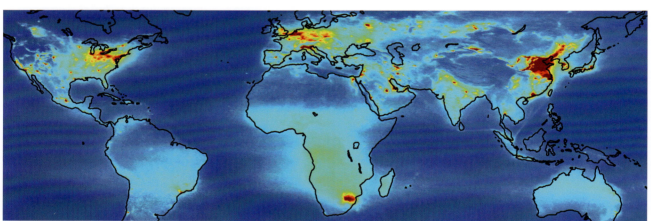

1 Sauberes Wasser führt zu Verbesserungen von Gesundheit, Bildung, Lebensunterhalt, Rolle der Frauen und Umwelt. *Clean water brings improvements in health, education, livelihoods, the role of women and the environment.* ▬▬ 2 20 Nanometer klein sind die Poren der Multibore®, durch die Wasser bei der Ultrafiltration gepresst wird, wie bei dieser Anlage in Roetgen in der Nordeifel. *The Multibore® pores are only 20 nanometres in size through which water is forced in the ultrafiltration process, as in this plant in Roetgen in the northern Eifel region.* ▬▬ 3 Luftverschmutzung anhand von Stickstoffdioxid-Messungen: Regionen mit hoher Verkehrs- und Industriedichte sind deutlich zu erkennen. *Air pollution based on nitrogen dioxide measurements: Regions with high transport and industry densities are clearly visible.* ▬▬ 4 Diese solarthermischen Meerwasserentsalzungsanlagen mit mehrstufiger Wärmerückgewinnung auf Gran Canaria erzeugten bei Feldtests etwa 200 Liter Trinkwasser pro Tag. *This solar thermal seawater desalination plant with multi-stage heat recovery in Gran Canaria produced around 200 litres of drinking water per day in field tests.* ▬▬ 5 In unverbauten Flüssen, wie hier in der unteren Donau in Bulgarien, können sich Fische und andere Lebewesen ungehindert im Gewässer bewegen. *In unobstructed rivers, such as the lower Danube in Bulgaria (pictured), fish and other animals can move unhindered through the water.*

1 SkyJuice Foundation 2 Rheinisch-Westfälisches Institut für Wasser IWW, Mülheim an der Ruhr; inge watertechnology AG 3 Steffen Beirle, Max-Planck-Institut für Chemie, Mainz; Daten: Instrument SCIAMACHY an Bord des ESA-Satelliten ENVISAT, 2003–2006 4 Solar-Institut Jülich, Hochschule Aachen; Ingenieurbüro für Energie und Umwelttechnik (IBEU), Jülich 5 »Rivers of Europe«, Tockner et al. 2009, Leibniz-Institut für Gewässerökologie und Binnenfischerei, Berlin

energies to the power consumption in Germany rose from 4.5 to 14.2 percent between 1997 and 2007. This figure is set to rise even further. However, renewable energies are not without their problems when it comes to supplying power. The energy they generate depends on the time of day and the weather, and is difficult to store for use at a later date. In the future, intelligent power grids, in which distributed capacities are interconnected to form virtual power plants, will help to solve these problems. ▬▬ **The sun in a reactor** Within a few decades, nuclear fusion could be enriching the energy mix as an environmentally friendly and safe energy source. The ITER fusion experiment is expected to release ten times the amount of energy originally needed to heat the plasma. Specially shaped magnetic fields have to be used to keep the plasma at the extreme temperatures – even higher than those inside the sun. ▬▬ **Electricity and heat from the sun** In thermal and photovoltaic plants, solar energy is already being used directly to provide us with warm water and electricity. New materials and technical developments are making it ever simpler to use solar energy plants – with increasing efficiency. A number of large-scale solar thermal plants for electricity generation are already in the planning or construction stages. ▬▬ **What about bioelectricity and biofuel?** Agricultural areas must be reserved for the cultivation of foodstuffs – scientists can now produce bioenergy from biological waste materials. Energy-rich plants used to produce bioethanol have a negative impact on the climate due to the fertilisers they require. Second and third generation techniques are capable of recovering bioenergy from waste wood, straw and even biological kitchen waste. ▬▬ **Energy from hydrogen?** Storable in principle and clean when burned: Hydrogen would be the ideal energy source. Efficient fuel cells already exist. But storage remains a problem: Compression, liquefaction and transport consume enormous amounts of energy and materials. Scientists are therefore redoubling their efforts in the search for chemical storage methods with liquid or solid carriers. ▬▬ **Using energy efficiently** ▬▬ The cheapest and most environmentally friendly energy is the energy that we don't consume. *Carlo Rubbia, Nobel Prize for Physics, 1984* ▬▬ Energy savings are possible in many areas: not only with more efficient machines and appliances, whether at home or in industry, but also in the spheres of power generation and transport. ▬▬ **Humankind is changing the ecological balance** Clean water, clean air and unspoiled natural habitats should be the goals for a sustainable use of our Earth. Maintaining the status quo is no longer enough to protect the basis of human life for the future.

Small-scale approaches to environmental protection show substantial promise. More than ever before, people are fighting pollution and the effects of climate change at a local level. Global actions to offset the greenhouse effect, such as geo-engineering, are hard to realise and usually come with serious side-effects. Reducing the amount of sunlight hitting the Earth or fertilising specific areas in the oceans to store CO_2 could have unforeseen effects on the global climate. **Clean water means life** More than one billion people on Earth have no access to clean drinking water. Small projects often have a big impact. Distributed and simple solutions for sewage treatment can produce not only clean drinking water but often yield valuable fertilisers, too. Much waste water is polluted with phosphate – a raw material whose stores will be exhausted within a few decades. **Small pores for clean air** Nitrogen dioxide, particulates and ozone are the principal causes of the air pollution from which many countries still suffer. Exhaust fumes originate mainly from power stations, heavy industry and transport. Better filters and more advanced industrial plants can help to keep damaging gases out of our environment or to convert them to relatively harmless materials. **Creating and sustaining biospheres** Biodiversity is not just important in the rainforest; there are ecosystems worthy of protection in our own backyard, too. In Europe, humankind changed the landscape centuries ago – to an extent that virtually none of the original ecosystems now remain. New ecosystems have since developed in our man-made environment and are exhibiting an astonishing diversity of species.

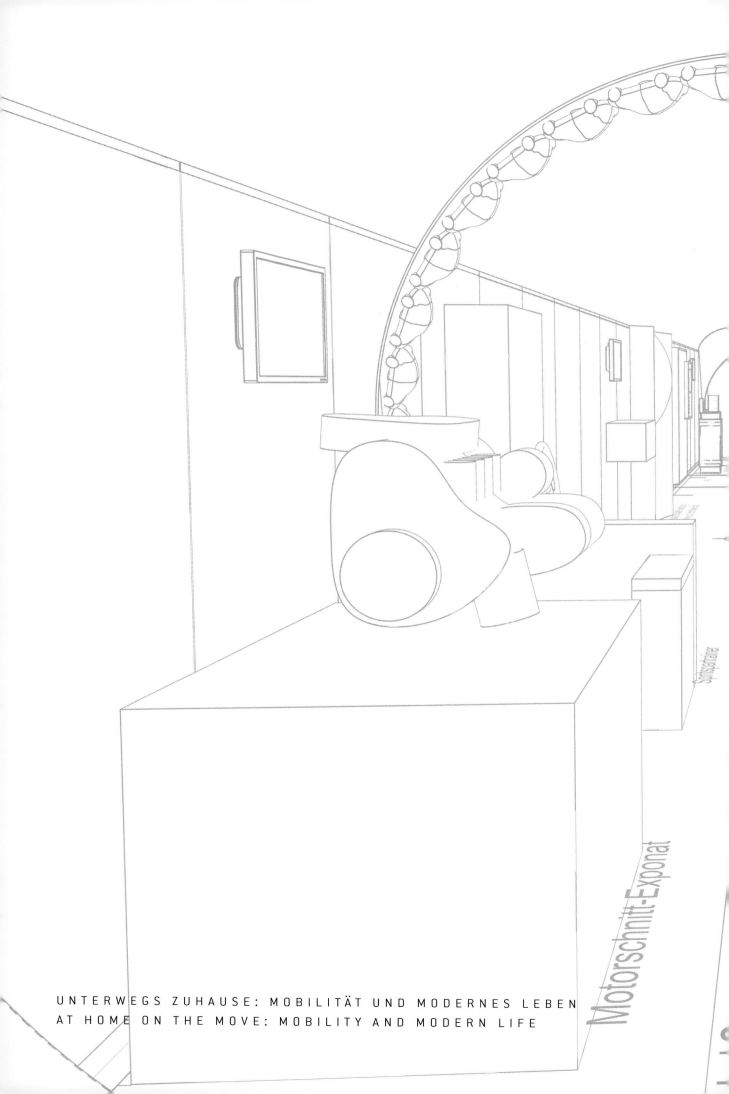

UNTERWEGS ZUHAUSE: MOBILITÄT UND MODERNES LEBEN
AT HOME ON THE MOVE: MOBILITY AND MODERN LIFE

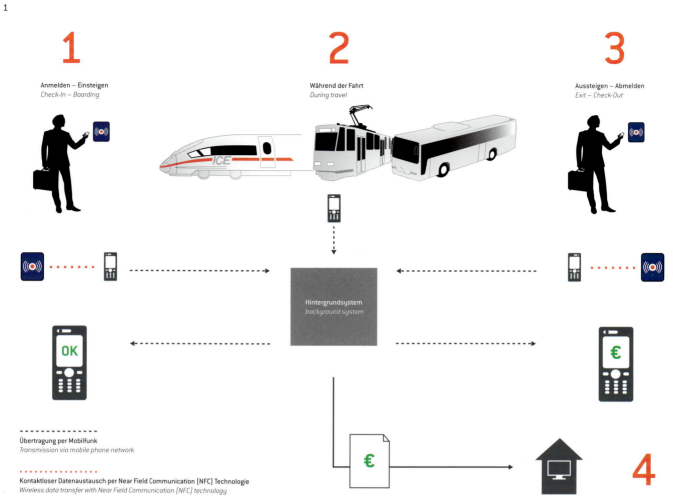

1 Anmelden – Einsteigen
Check-In – Boarding

2 Während der Fahrt
During travel

3 Aussteigen – Abmelden
Exit – Check-Out

Hintergrundsystem
background system

Übertragung per Mobilfunk
Transmission via mobile phone network

Kontaktloser Datenaustausch per Near Field Communication (NFC) Technologie
Wireless data transfer with Near Field Communication (NFC) technology

4

Unterwegs zu Hause: Mobilität und modernes Leben

Unsere Gesellschaft wird immer urbaner: Mehr und mehr Menschen leben und arbeiten in immer größeren Städten. Vor allem Schwellenländer stehen vor der Herausforderung, diese Megastädte angemessen und umweltverträglich mit Gütern, Infrastruktur und Wohnungen zu versorgen. Technische Entwicklungen zielen einerseits darauf ab, Fahrzeuge, Gebäude und Geräte energiesparender und umweltfreundlicher zu gestalten. Andererseits werden immer mehr Komponenten miteinander vernetzt und durch intelligente Systeme einfacher zu bedienen: Sie passen sich flexibel den Bedürfnissen der Benutzer an und bieten lückenlosen Zugriff auf Inhalte und Dienste. Durch intelligente Planung und Design können unnötige Wege vermieden, Energie gespart und Naturressourcen geschont werden. _____ **Auf den eigenen vier Rädern in die Zukunft** *Das Auto ist und bleibt das zentrale Fortbewegungsmittel – auch in Zeiten hoher Spritpreise. Seine Vorteile: zeitliche Flexibilität und ein flächendeckendes Straßennetz. Über 90 Prozent der Deutschen über 18 Jahre besitzen einen Führerschein. Innovationen in der Automobiltechnik machen Fahrzeuge sicherer, sparsamer und umweltfreundlicher. Bessere Motoren verbrauchen weniger Treibstoff, Katalysatoren reinigen Abgase und Sensoren unterstützen und überwachen alles, von den Rädern bis zum Wimpernschlag des Fahrers. Wie lange wird es noch dauern, bis unsere Autos eigenständig fahren und sich im Dialog mit anderen Fahrzeugen selbst den besten, schnellsten und sichersten Weg suchen?* _____ **Neue Energie unter der Haube** Fast alle Autohersteller arbeiten an neuen Antriebskonzepten; verbrauchs- und schadstoffarme Autos sind seit einigen Jahren auf dem Markt. Brennstoffzellen, Hybrid- und Elektroantriebe ergänzen den herkömmlichen Verbrennungsmotor, der auch biologisch erzeugte Kraftstoffe nutzen kann. Die Infrastruktur für diese Lösungen muss jedoch noch ausgebaut werden, damit sie sich im Alltagsverkehr durchsetzen können. _____ **Leichter und sicherer ans Ziel** Autos sind heute viel mehr als Stahlkisten mit Motor und Reifen: Maßgeschneiderte Kunststoffe und innovative Metallbauteile sind so leicht wie möglich und so stabil wie nötig. Werkstoffinnovationen tragen zur passiven Sicherheit im Auto bei und helfen, den Treibstoffverbrauch zu senken. Eine hohe Lebensdauer ist außerdem ein wichtiges Kriterium

Vorige Doppelseite: Designstudien für die Fortbewegung und Wohnkonzepte der Zukunft *Preceding spread: Design studies for future mobility and living concepts.* _____ 1 Bahnfahren mit Touch&Travel: Das NFC-Handy als eTicket. Das Ticket wird durch An- und Abmelden am Touchpoint automatisch erzeugt. Der Preis wird nach Fahrtende im Hintergrundsystem berechnet. *Train journeys with Touch&Travel: Automatic tickets via NFC mobile phones. The fare is calculated automatically by the background system based on the check-in and check-out with the mobile phone at the Touchpoint.* _____ 2 Moderne Motoren verbinden maximale Leistung mit sehr niedrigem Verbrauch und geringen CO$_2$-Emissionen. Der kompakte TSI-Motor beispielsweise besitzt deshalb einen Kompressor und einen Turbolader. *Modern engines combine maximum performance with very low fuel consumption and few CO$_2$ emissions. The compact TSI engine for instance includes a compressor and a turbocharger.* _____ 3 Weniger Spritverbrauch spart Geld und senkt die CO$_2$-Emissionen. Viele Spartipps, wie schnelles Hochschalten, lassen sich ganz einfach umsetzen. *Less fuel consumption saves money and lowers CO$_2$ emissions. Many suggestions on how to save fuel, such as quickly changing to higher gears, are easy to implement.* _____ 4 Konzeptstudie Hochgeschwindigkeitszug *Concept study of a high speed trainset*

Vorige Doppelseite *Previous spread* Siemens AG 1 Deutsche Bahn AG 2, 3 Leihgeber *On loan by* Volkswagen AG 4 Siemens AG

bei der Entwicklung von neuen, sich selbst reparierenden Lackierungen.
Clevere Autos melden sich zu Wort Schlaue Sensoren machen das Fahren nicht nur einfacher, sondern auch sicherer. Sie sorgen für ein effizientes Energiemanagement und warnen vor Gefahren. Intelligente Systeme erfassen die Situation in und um das Auto und machen den Fahrer auf mögliche Gefahren aufmerksam. Bald werden Fahrzeuge miteinander kommunizieren und sich gegenseitig vor Stau, Regen oder schlechten Straßenverhältnissen warnen. **Mobilität 2020: Der schnellste Weg von A nach B** *Das Verkehrsaufkommen wird in den kommenden Jahren weiter zunehmen; die zunehmende Vernetzung der Weltwirtschaft und die verstärkte Arbeitsteilung verlangen weltweit nach mehr Mobilität von Personen und Gütern. Durch intelligente Systeme und lückenlose Überwachung werden Waren optimal auf die vorhandenen Transportwege aufgeteilt und kommen genau zur rechten Zeit am Bestimmungsort an. Die Basis hierfür, eine vernetzte Infrastruktur und integrierte Informationssysteme, erleichtern auch uns das Reisen. Bei Bus und Bahn entfällt der Kauf von Fahrkarten; bei der Fahrt mit dem Auto können Staus schon im Vorfeld vermieden werden. Können wir so Mobilität und Umweltschutz in die richtige Balance bringen?*
Signale aus dem All lenken den Verkehr Ab 2013 soll das zivile Satelliten-Navigationssystem ›Galileo‹ Signale liefern, die eine zehnmal genauere Positionsbestimmung als mit GPS ermöglichen. Der kostenlose Basisdienst bietet neben einer hohen Zuverlässigkeit auch die Möglichkeit, Satellitensignale mit anderen Funksystemen zu kombinieren. Autofahrer könnten so minutengenau über die aktuelle Verkehrslage informiert werden. **Reisen im Schlaf** Reisegeschwindigkeit bis zu 320 km/h, Weltrekord 575 km/h. Ein modernes Eisenbahnwesen, vernetzt mit anderen Verkehrsträgern, schafft neue Möglichkeiten. Die Eisenbahn konnte in den letzten Jahren gegenüber Auto und Flugzeug aufholen. Neben schnellen Verbindungen und einfachen Buchungsmöglichkeiten werden andere Kriterien immer wichtiger, wie die Zeit zum Arbeiten oder Entspannen während der Fahrt. **Logistik wird intelligent** Güter, die mit Chips zur einfachen Identifikation durch Radiowellen ausgestattet sind, können automatisch lokalisiert und überwacht werden. So kann etwa die Kühlkette lückenlos dokumentiert sowie die Herkunft und Frische der Waren sichergestellt werden. Darüber hinaus könnten Container mit RFID-Chips künftig eigenständig ihren Weg finden und Störungen schon unterwegs melden. **Millionen in der Stadt der Zukunft** *Seit 2008 lebt zum ersten Mal in der Geschichte die Mehrheit der Menschheit in Städten. Die meisten und am schnellsten wachsenden Megastädte mit mehr als zehn Millionen Einwohnern findet man vor allem in Schwellen- und sich entwickelnden Ländern. Das explosionsartige Wachstum vieler Städte stellt die Stadtplaner vor neue Herausforderungen, technische wie soziale. Nach der Grundversorgung mit Wohnraum, Essen und Wasser, Sanitäranlagen und Strom steht Lebensqualität an erster Stelle. Die Menschen wollen schnell und problemlos zur Arbeit oder zum Arzt gelangen und flexibel ihre Besorgungen erledigen können. Gleichzeitig gewinnt Nachhaltigkeit immer mehr an Bedeutung. Dies verleiht der Verkehrsplanung und Bautechnik neue Impulse.* **Bessere Chancen in Metropolen** Mit mehr als zehn Millionen Einwohnern ist die Rhein-Ruhr-Region die einzige deutsche Megastadt. Die meisten dieser Millionenstädte liegen in Asien und Südamerika. Slums, Abfallberge, Smog – viele Megastädte stehen vor großen sozialen, ökologischen und ökonomischen Herausforderungen. In vielen Teilen der Welt sind Städte trotzdem attraktiv: Sie bieten bessere Chancen auf Arbeit und einen gewissen Lebensstandard. **Hightech verändert den Bau** Aufgrund steigender Heizkosten und höherem Umweltbewusstsein ist die Energieeffizienz bei Neu- und Umbauten inzwischen eines der wichtigsten Kriterien. Statt Wärme durch schlechte Dämmung zu verlieren, können Mauern und Fenster sogar zur Energieerzeugung genutzt werden. Neue Beschichtungen ›veredeln‹ Wände,

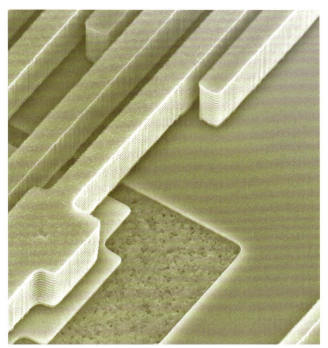

Fenster und Böden. **Unser Leben wird digital** *Eine Wohnung ist heute mehr als nur ein Dach über dem Kopf. Viele Geräte erleichtern uns den Alltag und sorgen für Sicherheit, Behaglichkeit und Unterhaltung. Intelligentes Management hilft, Energie und Zeit zu sparen. Die moderne Haustechnik muss dabei unterschiedlichste Anforderungen erfüllen: Einbauten und Geräte sollen einfach zu bedienen sein und sich dezentral und flexibel an äußere Bedingungen und unterschiedliche Raumnutzung anpassen. Grundlage hierfür sind neue Entwicklungen in der Mikroelektronik, der drahtlosen Datenübertragung und der Softwaretechnik. Die größte Herausforderung besteht darin, verschiedene Systeme nahtlos zu verbinden und so neuen Service im eigenen Haus zu schaffen. Wann wird das vernetzte Heim Wirklichkeit?* **Licht ist mehr als Helligkeit** Moderne Leuchten verbrauchen sehr viel weniger Strom und produzieren dabei ein brillanteres und natürlicheres Licht als die gute alte Glühbirne. Es gilt Abschied zu nehmen. Intelligente Lichtinstallationen passen sich individuell unseren Bedürfnissen und Stimmungen an. Wonach steht Ihnen der Sinn: Tageslicht zum konzentrierten Arbeiten oder ein sanftes Licht zum Wohlfühlen? Licht selbst, nicht die Lampe, wird zum Designelement. **Zu Hause in einer digitalen Welt** Intelligente Dienste verbinden künftig Geräte von der Waschmaschine bis zu den Jalousien, passen sich individuell dem Bewohner an und senken Energiekosten. Das Haus der Zukunft ist mit der Welt vernetzt: Die Nutzer können überall auf ihre persönlichen Informationen und Dienste zugreifen. Das erleichtert die Haushaltsführung, steigert das Wohlbefinden und eröffnet neue Möglichkeiten bei der Kommunikation und für die Unterhaltungselektronik. **Sensoren werden intelligent** Mobile Geräte können sich vielleicht bald selbst mit Energie versorgen: aus Vibrationen, Temperaturunterschieden oder Luftströmungen. Bald könnte auch die Bedienung einfacher werden: Intelligente Sensoren überwachen das Umfeld und reagieren automatisch auf bestimmte Reize wie beispielsweise Bewegungen. Menüs und Befehle zu lernen könnte damit bald der Vergangenheit angehören.

1 Im Fluoreszenzkonzentrator filtern farbige Platten den hochenergetischen Teil des Sonnenlichts und lenken diesen auf Solarzellen am Plattenrand. *The high-energy component of sunlight is filtered in the coloured panes of the fluorescence concentrator and directed to solar cells on the edges.* 2 **Vernetzes Leben** *Connected Life*

1 Leihgeber *On loan* by Fraunhofer Institut für Solare Energiesysteme ISE, Freiburg 2 Leihgeber *On loan* by Deutsche Telekom AG, Technische Universität Berlin

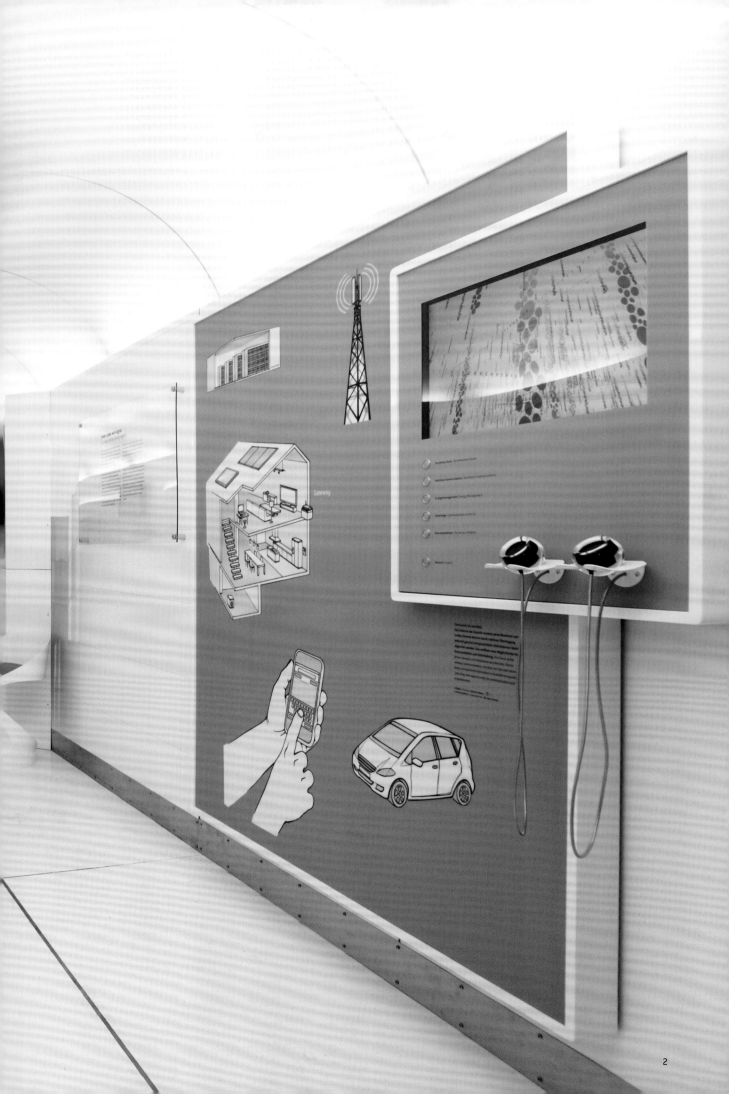

At home on the move: Mobility and modern life

Our society is becoming ever more urbanised: More and more people live and work in cities that are growing steadily larger. The newly industrialising countries in particular are faced with the challenge of ensuring that these megacities are adequately provided with goods, infrastructure and housing, in an environmentally sustainable manner. Technical developments are directed, on the one hand, towards designing vehicles, buildings and appliances that use less energy and are more environmentally friendly. On the other hand, more and more components are being networked together and becoming simpler to operate with the aid of intelligent systems: They adapt themselves to the needs of users, offering seamless access to content and services. Intelligent planning and design enable us to avoid unnecessary journeys, save energy and conserve natural resources.

Heading for the future on your own four wheels *The car is and remains our principal means of transport – even in times of high fuel prices. It offers the advantages of flexibility and a ubiquitous road network that we can use any time. More than 90 percent of Germans over the age of 18 have a driver's license. Innovations in automobile technology make cars safer, more economical and more environmentally friendly. Improved engines use less fuel, catalytic converters clean up exhaust fumes, and sensors support and monitor everything from the wheels to the blink of the driver's eye. How long will it be before our cars drive themselves and communicate with other vehicles to decide which route is the best, fastest and safest?*

New energy under the bonnet Almost all motor manufacturers are working on new drive system concepts. Low-consumption, low-emission cars have been on the market for some years now. Fuel cells, hybrid and electric drives complement the conventional combustion engine, which can also use biologically produced fuels. However, the infrastructure to support these solutions must still be expanded for them to establish themselves in day-to-day use.

Getting there – lighter and safer *Cars nowadays are much more than steel boxes with an engine and tyres. Customised plastics and innovative metal components are as light as possible, and as strong as need be. Innovations in materials contribute to passive in-car safety and help to reduce fuel consumption. Longevity is also an important criterion in the development of new, self-repairing paint finishes.*

Clever cars have a word to say Intelligent sensors make driving not just easier, but safer too. They guarantee efficient energy management and warn of dangers. Intelligent systems grasp the situation in and around the vehicle and draw the attention of the driver to possible hazards. Soon, vehicles will be able to communicate with one another and warn each other of traffic jams, rain or bad road conditions.

Mobility 2020: The quickest way from A to B *Traffic volumes will continue to increase in the coming years. The increasing integration of the global economy and the intensified division of labour demand greater mobility of people and goods the world over. With the aid of intelligent systems and seamless monitoring, goods can be allocated to the best transport routes available to ensure they arrive at their destination right on time. The underlying technology, a networked infrastructure and integrated information systems, makes travelling easier for us as well. There is no longer any need to buy tickets for bus or train. And when travelling by car, hold-ups can be averted preemptively. Is this the way to strike the right balance between mobility and environmental protection?*

Signals from space direct traffic From 2013 onwards, the civilian satellite navigation system ›Galileo‹ is scheduled to provide signals that will enable us to fix our position with ten times greater accuracy than is possible using GPS. Besides being highly reliable, the free basic service also offers the facility to combine satellite signals with other radio systems. Drivers will be able to receive up-to-the-minute details on the current traffic situation.

Travelling in your sleep *Cruising speeds of up to 320 km/h. A world record of 575 km/h. Modern railway systems integrated with other modes*

Die Megastadt Shanghai. Jede Megastadt hat ihre eigenen Probleme, einige betreffen aber alle: Stromversorgung, Mobilität, umweltverträglich neuen Wohnraum schaffen. *Megacity Shanghai. Each megacity faces its very own challenges, but some are common to all: energy, mobility, creating environmentally friendly new living space.*

Siemens AG

Städtische Ballungsräume 2025 *Population of urban agglomeration 2025*

NASA, mit Daten aus ›World Urbanization Prospects – The 2007 Revision‹, UN, New York

- Los Angeles
- New York
- Mexico City
- São Paulo
- Rio de Janeiro
- Buenos Aires
- Lagos

Population in Millionen
Population in million

>30 | 20–30 | 15–20 | 12–15

of transport create new possibilities. In recent years, railways have made up much lost ground with respect to cars and planes. In addition to fast connections and simple booking facilities, other criteria are gaining in importance, such as time to work or relax during the journey.⎯⎯**Adding intelligence to logistics** Goods that are fitted with chips for easy identification by radio frequency can be automatically located and monitored. This allows, for example, the cool chain to be seamlessly documented and the origin and freshness of the goods to be assured. In future, containers fitted with RFID chips may well be able to find their own way to their destination and report any incidents in transit.⎯⎯**The city of the future will be home to millions** *Since 2008, the majority of mankind lives in cities – for the first time in history. Most, and certainly the fastest growing of the megacities with more than ten million inhabitants are to be found in the newly industrialising and developing countries. The explosive growth experienced by many cities presents new challenges, both technical and social, for urban planners. Along with the basic provision of accommodation, food and water, sanitation and electricity, quality of life is at the top of the list. People want to get to work or to the doctor easily, and run their errands flexibly. At the same time, sustainability is becoming increasingly important, adding a new dynamic to traffic management and building technology.*⎯⎯**Opportunities are better in the metropolis** With more than ten million inhabitants, the Rhine-Ruhr region is Germany's only megacity. Most of these cities with millions of inhabitants are to be found in Asia and South America. Slums, garbage mountains, smog – many megacities face huge social, ecological and economic challenges. Despite this, in many parts of the world cities are attractive: They offer better opportunities to find work and achieve a certain standard of living.⎯⎯**High-tech is changing the face of construction** As heating costs rise and environmental awareness grows, energy efficiency is now one of the major criteria in both new builds and conversions. Instead of losing heat through poorly insulated walls and windows, these can actually be used to create energy. New surface coatings now upgrade walls, windows and floors.⎯⎯**Our lives are becoming digital** *Today, more than ever, a*

Container und Containerschiffe – Symbol der Globalisierung. Mithilfe drahtloser Sensoren innerhalb moderner Container werden Temperatur, Erschütterung, Feuchte, Reifegase und andere Parameter gemessen und online bewertet. So kann etwa der ›intelligente Container‹ der Universität Bremen gezielt warnen und sogar auf Probleme reagieren. *Containers and container ships – symbols of globalization. Using wireless sensors, the ›intelligent container‹ (University of Bremen) measures temperature, shocks, humidity, ripeness gases and other parameters and evaluates these online. It can thus give specific warnings and even react to problems.*

Siemens AG

1 Höhere Sicherheit beim Autofahren steht bei der Entwicklung neuer Sensoren im Mittelpunkt: Elektronische Fahrerassistenzsysteme können zur Unfallvermeidung und Minderung von Unfallfolgen beitragen. Sie erkennen Gefahren, geben rechtzeitig Warnungen oder greifen selbst aktiv ein. *Increased security in the car is the main goal when developing new sensors: Electronic driver assistance systems can help to prevent accidents or alleviate their effects. They detect dangerous situations, give early warnings or intervene themselves.* —— 2 Moderne LEDs in Autoscheinwerfern liefern helles, weißes Tagfahrlicht bei geringem Energieverbrauch. Die roten LEDs am Heck reagieren schneller und verbrauchen rund zehnmal weniger Energie als herkömmliche Bremslichter. *The modern LEDs used in vehicle headlights produce bright, white light with low power consumption. The red LED at the rear reacts quicker and uses about ten times less energy than conventional break lights.* —— 3   Visionen zum Auto der Zukunft *Visions for the car of the future*

1 Leihgeber *On loan by* Volkswagen AG   2 Leihgeber *On loan by* OSRAM GmbH   3 Volkswagen AG

home is more than a roof over our heads. A host of appliances make life easier, providing security, comfort and entertainment. Intelligent management helps to save energy and time. Modern technical services must satisfy a wide variety of needs: Fixed and portable appliances must be simple to operate and adapt themselves decentrally and flexibly to external conditions and changes in how we use our living space. New developments in microelectronics, wireless data transmission and software technology offer a starting point. The biggest challenge is to seamlessly interconnect a variety of systems in order to create a new level of service in our own homes. When will the networked home become a reality?

Light is more than luminosity Modern lamps use much less electricity and produce a more brilliant and natural light than the good old light bulb. Time to say goodbye. Intelligent lighting installations adapt themselves individually to our needs and moods. Which would you prefer: daylight to concentrate on work, or a gentle illumination for that feel-good factor? Light itself has replaced the lamp as a design element. **At home in a digital world** Intelligent services will, in future, link appliances – from the washing machine to the window blinds – adapting themselves individually to the occupants, and lowering energy costs. The house of the future will be networked with the world: Users will be able to access personalised data and services, wherever they may be. This, in turn, will make house and home easier to manage, increase well-being and open up new opportunities in communication and entertainment electronics. **Sensors are acquiring intelligence** Mobile devices may soon be able to supply themselves with energy: From vibrations, temperature differences or airflows. Operation, too, could soon become easier: Intelligent sensors monitor their surroundings and respond automatically to certain stimuli, such as movements. The need to learn menus and commands could soon be a thing of the past.

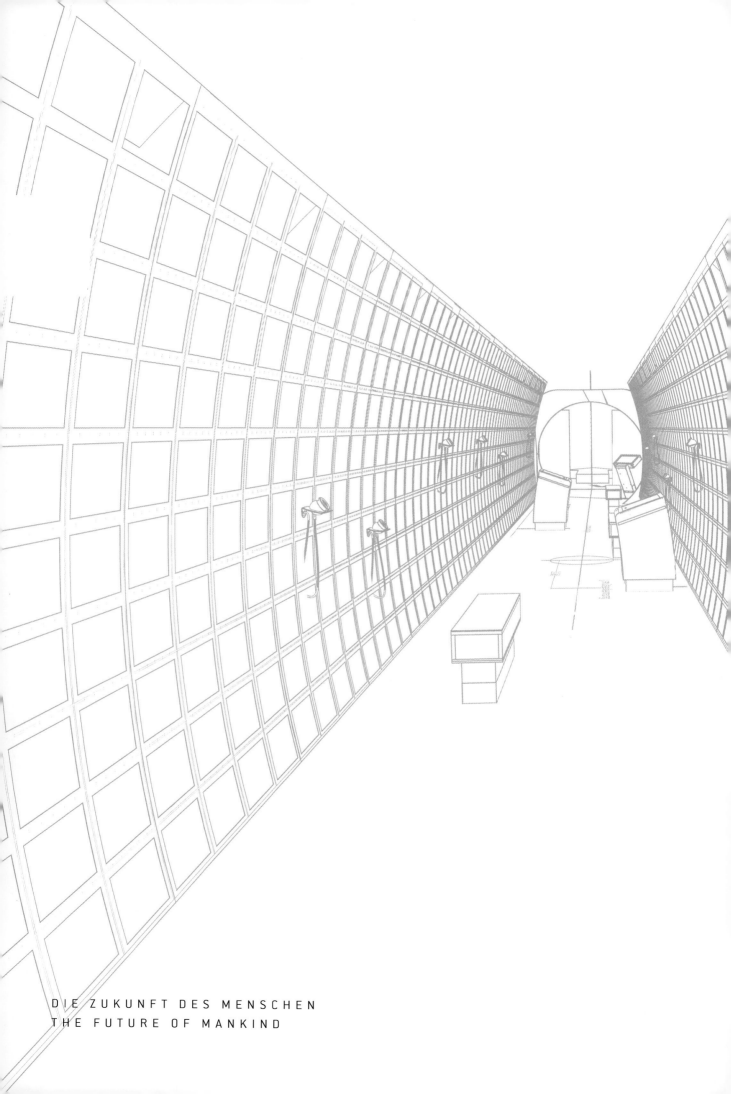

DIE ZUKUNFT DES MENSCHEN
THE FUTURE OF MANKIND

natürlich. künstlich

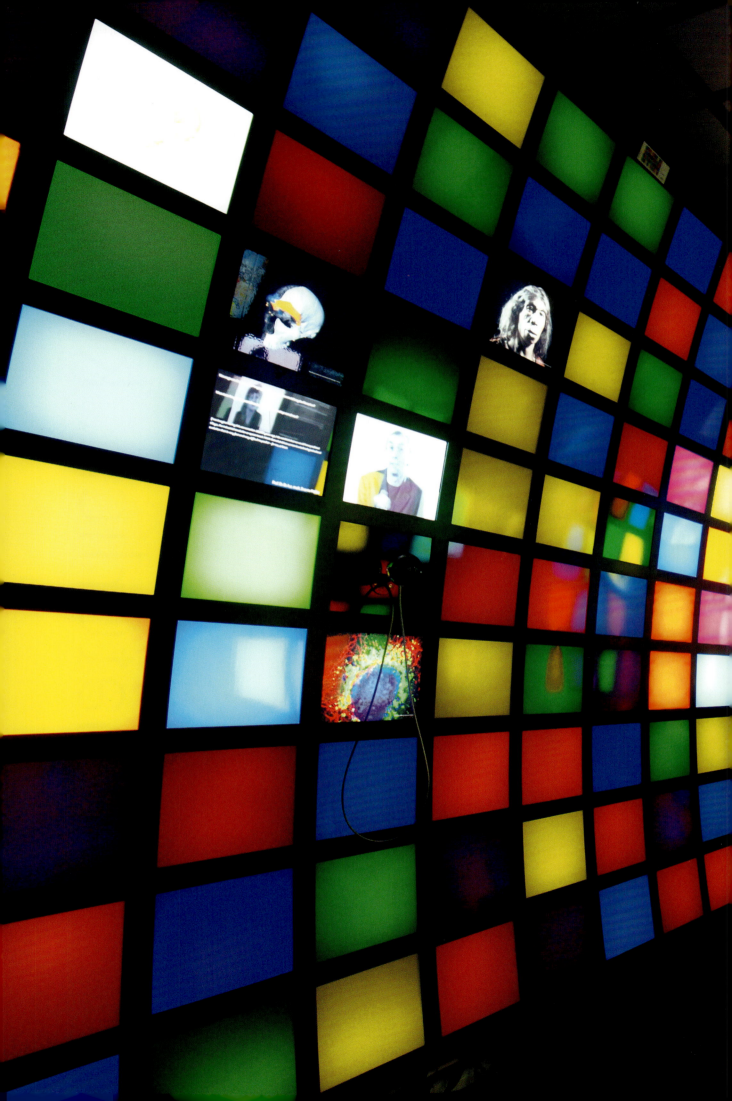

მომავალი ადამიანი
Zukunft Mensch

未来人

Будущее человек

انسان آینده

[tsuk

ভবিষ্য মানুষ

مستقبل انسان
انسان آینده

مستقبل انسان

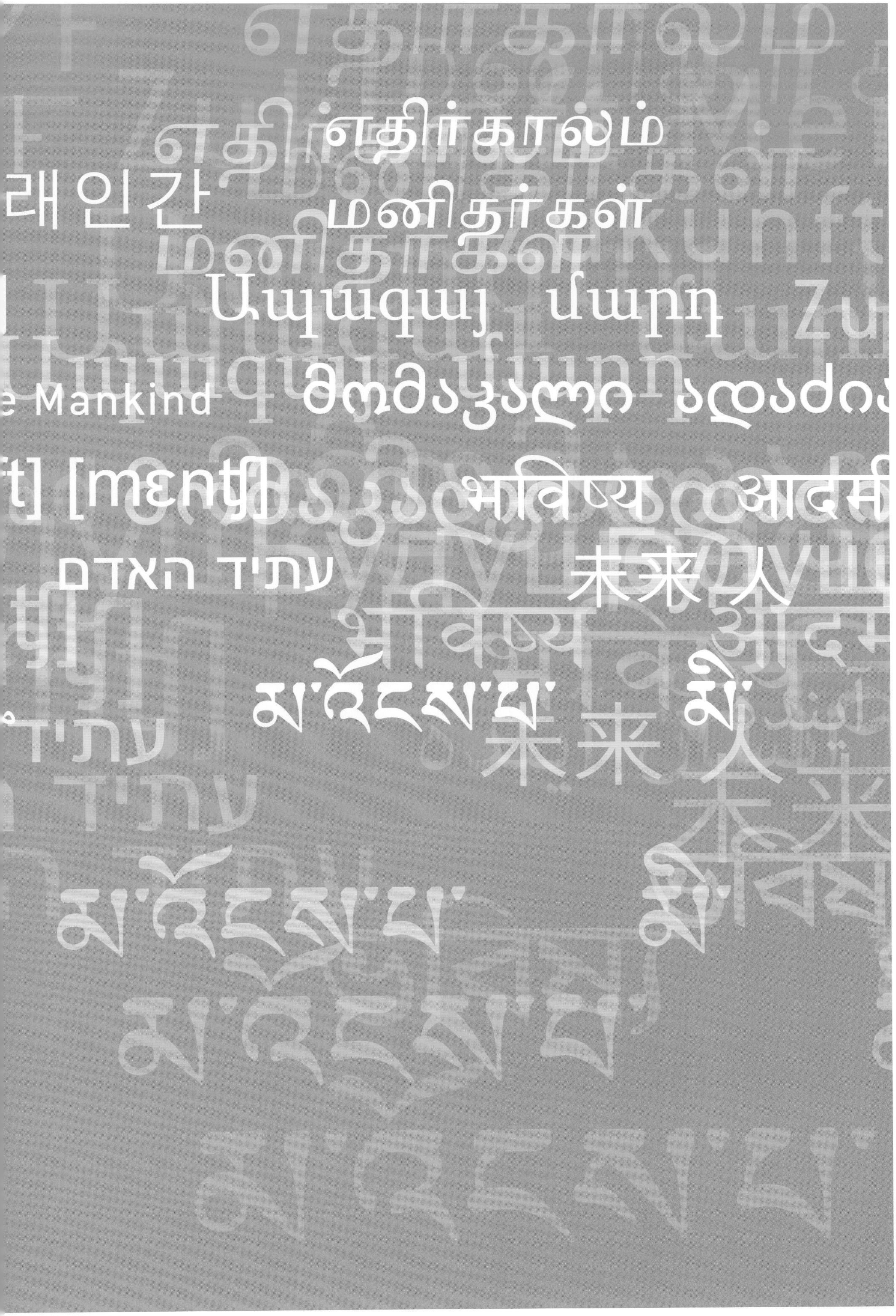

எதிர்காலம் மனிதர்கள்

Ապագայ մարդ

მომავალი ადამი

भविष्य आदर्श

未来

Mankind

[mɛntji]

עתיד האדם

མ་འོངས་པ་

미래 인간

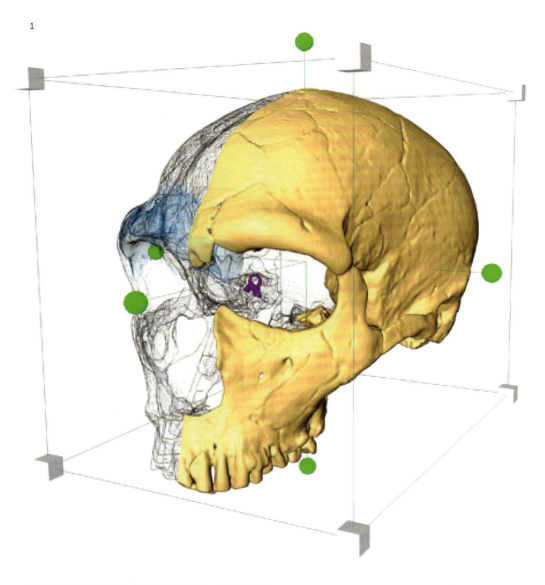
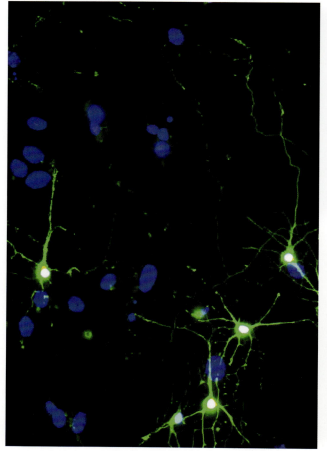
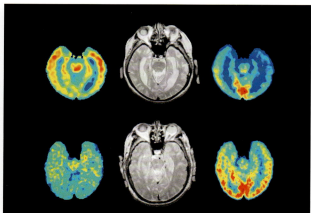

Die Zukunft des Menschen

Das menschliche Bewusstsein ist das Ergebnis natürlicher Evolution. Der Mensch aber setzt diese Evolution mit künstlichen Mitteln der Wissenschaft und Technik fort. Gene, Neurone, Meme, das Lachen und ›Zeigen‹ erneuern die Perspektiven unserer Selbsterkenntnis. Genetik, Kognitionsforschung und Informationstechnologien greifen in die künftige Entwicklung des Menschen ein; Wissenserwerb und Bildung des Einzelnen, aber auch der Gesellschaft, werden sich ändern. Gentechnik trägt zur Klärung der Menschheitsgeschichte bei und stellt bald jedem das Wissen über sein Genom zur Verfügung. Die Reproduktionsmedizin verhilft Kinderlosen zu Nachwuchs. Der Mensch bedient sich intelligenter technischer Artefakte, um natürliche Grenzen zu erweitern. Wo aber sind die ethischen Grenzen und welche Möglichkeiten bietet die Zukunft? Wissenschaftler, Gelehrte und Künstler geben Antworten auf die zentrale Frage: ›Was ist der Mensch?‹ **Prof. Dr. Svante Pääbo, Paläoanthropologe, Max-Planck-Institut für evolutionäre Anthropologie, Leipzig:** ›Wenn wir das Genom des Neandertalers mit dem des Menschen vergleichen, dann hoffen wir die genetischen Grundlagen zu finden, die den Menschen einzigartig machen. Was in unserer Biologie versetzt uns in die Lage, Technologie zu entwickeln, Sprache zu entwickeln, Kunst zu entwickeln und die ganze Welt zu kolonisieren?‹ **Prof. Dr. Hans Lehrach, Molekulargenetiker, Max-Planck-Institut für Molekulare Genetik, Berlin:** ›Vor allem in der Krebsbehandlung sind wir, glaube ich, jetzt schon in der Lage, die Kenntnis des Genoms, des Tumors und des Patienten zu verwenden, um sehr viel gezielter die vorhandenen Behandlungen einzusetzen, [das] kann aber auch in vielen anderen Bereichen des Lebens wichtig werden. Die Abschätzung, welche Mittel gegen Haarausfall ein bestimmter Mensch verwenden sollte, oder die Optimierung des Trainings im Fitness-Studio aufgrund des individuellen Genoms würden mich ebenfalls nicht überraschen.‹ **Dr. med. Monika Bals-Pratsch, Reproduktionsmedizinerin, Universität Regensburg:** ›Die künstliche Befruchtung wird sicher die natürliche Zeugung nicht ablösen, aber es gibt viele Paare, die die Möglichkeiten der künstlichen Befruchtung nutzen müssen, um eigene Kinder zu bekommen, aus medizinischen Gründen, die sich dann für die Samenspende oder für die Eizellspende oder auch für die Leihmutterschaft entscheiden. Es gibt Samenbanken, es gibt Eizellbanken, und das bietet natürlich auch gleichgeschlechtlichen Paaren die Möglichkeit, eigene Kinder zu bekommen, eine eigene Familie zu gründen, eine so genannte Regenbogenfamilie.‹ **René Röspel, Biologe, Mitglied des deutschen Bundestags, Berlin:** ›Wie kaum ein anderes Land ist Deutschland abhängig von Grundlagenforschung und technischer Innovation und natürlich haben Innovation, Wissenschaft und Technik Einfluss auf die Gesellschaft und auf die Fragen von Ethik und Moral.‹ **Prof. Dr. Niels Birbaumer, Neuropsychologe, Universität Tübingen:** ›Ich glaube, dass schon in naher Zukunft der Mensch und das Gehirn des Menschen und ein angeschlossener Computer eine Art Symbiose eingehen können.‹ **Prof. Dr. Thomas Christaller, Robotikforscher, Universität Bielefeld und Institutsleiter am Campus Schloss Birlinghoven:** ›Ich finde eine wissenschaftliche Vision, die in der neueren Zeit definiert wurde, sehr interessant, nämlich, dass im Jahr 2050 ein Team von humanoiden Robotern in der Lage ist, den amtierenden Weltmeister im Fußballspielen potenziell schlagen zu können – eine große Herausforderung. Wenn das möglich ist – was offen ist – dann werden diese Roboter sicherlich vom Fairplay bis zur Blutgrätsche jedes Verhalten zeigen. Und sie werden sich sicherlich auch über die Schiedsrichterentscheidungen ärgern.‹ **Prof. Dr. Dr. Hannelore Ehrenreich, Medizinerin, Universität Göttingen und Max-Planck-Institut für Experimentelle Medizin, Göttingen:** ›Natürlich kommt die Frage auf, ob man Erythropoetin zum Hirndoping auch bei Gesunden einsetzen könnte. Dies kann ich als Wissenschaftlerin nicht befürworten. Ich gehe davon aus, dass die Stimulation

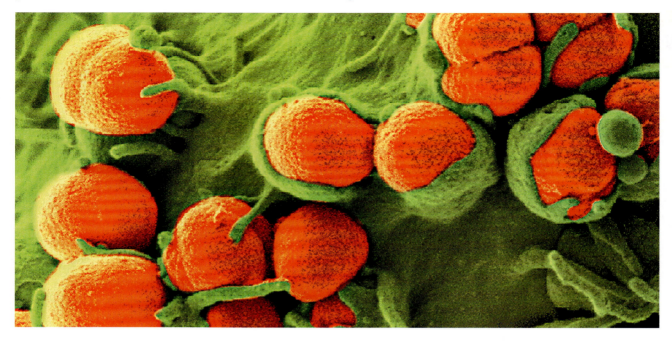
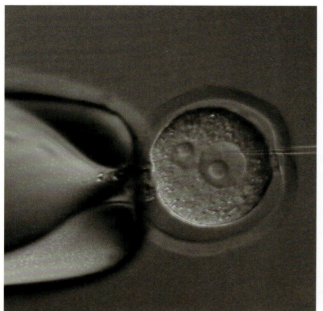
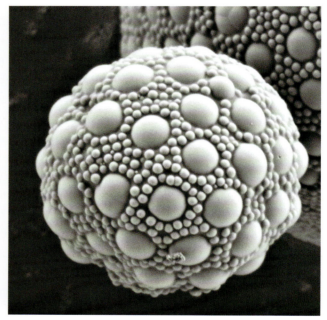
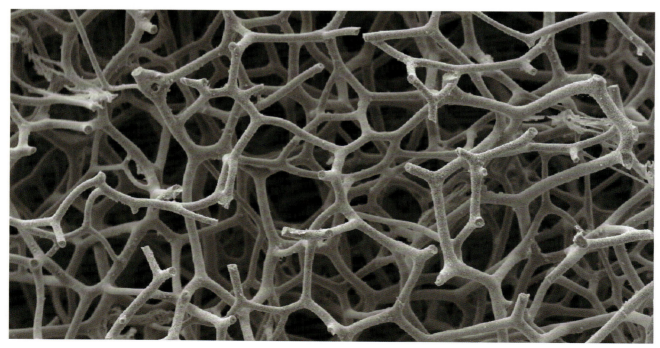

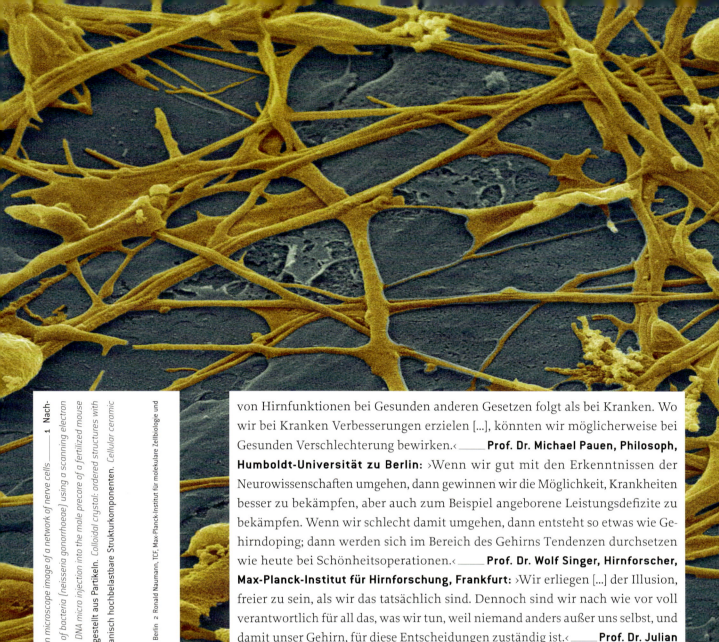

Hintergrund: Elektronenmikroskopische Aufnahme eines Netzwerks von Nervenzellen *Background: Electron microscope image of a network of nerve cells* ────── 1 Nachkolorierte Rasterelektronenmikroskop-Aufnahme von Bakterien (Neisseria gonorrhoeae). *Colorized image of bacteria (neisseria gonorrhoeae) using a scanning electron microscope.* ────── 2 DNS-Mikroinjektion in den männlichen Vorkern einer befruchteten Eizelle der Maus. *DNA micro injection into the male precore of a fertilized mouse egg cell.* ────── 3 Kolloidales Kristall: geordnete Strukturen mit besonderen optischen Eigenschaften hergestellt aus Partikeln. *Colloidal crystal: ordered structures with special optical properties made from particles.* ────── 4 Zellularer Keramikschaum für leichte aber mechanisch hochbelastbare Strukturkomponenten. *Cellular ceramic foam for lightweight but highly resilient structures.*

Hintergrund *Background* Jürgen Berger, Max-Planck-Institut für Entwicklungsbiologie, Tübingen 1 Max-Planck-Institut für Infektionsbiologie, Berlin 2 Ronald Naumann, TCF, Max-Planck-Institut für molekulare Zellbiologie und Genetik 3, 4 Exzellenzcluster ›Engineering of Advanced Materials‹, Erlangen

von Hirnfunktionen bei Gesunden anderen Gesetzen folgt als bei Kranken. Wo wir bei Kranken Verbesserungen erzielen […], könnten wir möglicherweise bei Gesunden Verschlechterung bewirken.‹ ────── **Prof. Dr. Michael Pauen, Philosoph, Humboldt-Universität zu Berlin:** ›Wenn wir gut mit den Erkenntnissen der Neurowissenschaften umgehen, dann gewinnen wir die Möglichkeit, Krankheiten besser zu bekämpfen, aber auch zum Beispiel angeborene Leistungsdefizite zu bekämpfen. Wenn wir schlecht damit umgehen, dann entsteht so etwas wie Gehirndoping; dann werden sich im Bereich des Gehirns Tendenzen durchsetzen wie heute bei Schönheitsoperationen.‹ ────── **Prof. Dr. Wolf Singer, Hirnforscher, Max-Planck-Institut für Hirnforschung, Frankfurt:** ›Wir erliegen […] der Illusion, freier zu sein, als wir das tatsächlich sind. Dennoch sind wir nach wie vor voll verantwortlich für all das, was wir tun, weil niemand anders außer uns selbst, und damit unser Gehirn, für diese Entscheidungen zuständig ist.‹ ────── **Prof. Dr. Julian Nida-Rümelin, Philosoph, Ludwig-Maximilians-Universität, München:** ›Was die Neurowissenschaft nicht zeigen kann ist, dass das Gründeabwägen keine Rolle spielt für unsere Überzeugungen und für unsere Handlungen. Wenn es so wäre, wäre eigentlich auch das Projekt der Wissenschaft obsolet.‹ ────── **Prof. Dr. Alfred Nordmann, Philosoph, Technische Universität Darmstadt:** ›Die Nanotechnologie zeichnet sich dadurch aus, dass die Dinge tatsächlich unheimlich, unvorstellbar klein sind. Das Unheimliche besteht darin, dass diese Technik aus unserem Erfahrungshorizont verschwindet und etwas quasi Magisches, etwas quasi Naturhaftes wird.‹ ────── **Prof. Dr. Gerd Karl Binnig, Physiker, Honorarprofessor der Ludwig-Maximilians-Universität, München:** ›Als wir vor mehr als 25 Jahren mit der Entwicklung der Rastertunnelmikroskopie angefangen haben, da erschien mir das relativ normal und schon fast real, mit den Atomen zu spielen. Wenn man an irgendetwas fest glaubt, dann wird es fast zur Realität. Aber wenn ich mir heute anschaue, mit welcher Selbstverständlichkeit die Leute mit den Atomen umgehen, mit dieser atomaren Welt umgehen, dann kommt mir das schon fast unwirklich vor. Also ich würde sagen, diese Selbstverständlichkeit ist die eigentliche Überraschung.‹ ────── **Prof. Dr. Dr. h. c. mult. Wolfgang Wahlster, Informatiker, Deutsches Forschungszentrum für Künstliche Intelligenz in Saarbrücken, Kaiserslautern, Bremen und Berlin, und Universität des Saarlandes:** ›Intelligente Softwaresysteme können die multimedialen Inhalte des Web immer besser inhaltlich verstehen, sodass quasi ein semantisches Web entsteht. […] Damit wird der Zugriff multimodal, der Computer wird zum Dialogpartner und der Zugang ist für alle ganz intuitiv.‹ —— **Prof. Dr. Gerhard Schulze, Soziologe, Universität Bamberg:** ›Wenn wir auf die mehrhundertjährige Erfolgsgeschichte der Moderne zurückblicken, müssen wir sagen, es hat wunderbar funktioniert. Allerdings hat es auch absurde Züge, denn über all diesem Steigern, dieser Vermehrung der Möglichkeiten, diesem Hinwollen zur besten aller Welten, sind wir immer auch unzufrieden und haben eine Fähigkeit verlernt, ohne die das Steigern völlig sinnlos ist, nämlich im gesteigerten Möglichkeitsraum uns aufzuhalten. Das heißt: Zeit zu haben, nicht mehr steigern zu wollen, sondern das, was man erreicht hat, ausnutzen zu wollen, sich daran zu erfreuen.‹ —— **Prof. Dr. Elsbeth Stern, Lernforscherin, ETH Zürich, Schweiz:** ›Die Tatsache, dass wir ohne ein Gehirn nicht lernen können, heißt nicht, dass Hirnforscher wirklich die Schule verbessern können. Das können sie derzeit überhaupt nicht. Es gibt kein einziges Ergebnis der Hirnforschung, das uns zum Umdenken in der Schule zwingen würde. Was die bisherige empirische Lernforschung aus der Pädagogik und der Psychologie hervorgebracht hat, das geht bei weitem über die trivialen Aussagen mancher Hirnforscher hinaus.‹ —— **Prof. Dr. Ursula M. Staudinger, Psychologin, Jacobs-Universität Bremen:** ›Das Lernen ist ein Prozess, der grundsätzlich mit der menschlichen Existenz verknüpft ist. Solange wir leben, können wir lernen, wenn wir keinen tief greifenden pathologischen Prozessen unterliegen.‹ —— **Prof. Dr. Aleida Assmann, Literaturwissenschaftlerin, Universität Konstanz:** ›Heute können wir überall, wo wir sind und wann wir wollen, alle Opern von Händel hören, alle Songs der Beatles. Wir haben Zugriff auf das kulturelle Gedächtnis, auf die Bestände und können sie auch selber sammeln. Das ist ein Effekt einer großen Demokratisierung. Aber die Sache hat auch einen Haken, denn diese neuen Medien sind nicht mehr so stabil und sie erfordern sehr hohen technischen Aufwand und Schwierigkeiten, um sie immer wieder zu erneuern.‹ —— **Prof. Dr. Dr. h. c. mult. Hermann Parzinger, Archäologe, Präsident der Stiftung Preußischer Kulturbesitz, Berlin:** ›In Europa und in vielen Teilen der Welt leben wir in einer multikulturellen Gesellschaft. Viele Menschen prägen das Leben in unseren Städten mit Migrantenhintergrund, sie bringen ihre Geschichte mit. Ich denke, es ist für uns sehr wichtig, diese Geschichte zur Kenntnis zu nehmen und ihre Kultur besser zu verstehen.‹ —— **Volker Schlöndorff, Regisseur, Potsdam:** ›Durch Hirnforschung, durch die Frage, ob wir überhaupt einen freien Willen haben, wird ja auch die klassische Dramaturgie in Frage gestellt. Weil, wenn ich keinen freien Willen habe, müssen viele Konflikte ganz anders gelöst werden, Geschichten müssen dann ganz anders erzählt werden.‹ —— **Prof. Dr. Carsten Niemitz, Humanbiologe, Freie Universität Berlin:** ›Das echte Lachen ist tatsächlich ein Ausdruck des Menschenwesens. Wir haben es in der Evolution ererbt. Menschenaffen können zwar lachen, aber wir haben in dem Lachen die Empathie, das Eindenken in den anderen beim gemeinsamen Amüsement, und wir haben außerdem in der Evolution noch hinzu erworben, das Auslachenkönnen, das Ausgrenzenkönnen. Wir haben die Möglichkeit, mit dem Lachen soziale Grenzen zu schaffen.‹ —— **Prof. Dr. Dieter Hattrup, Theologe, Universität Paderborn:** ›Der alte Gegensatz zwischen Wissenschaft und Religion ist für mich passé. Es kommt darauf an, wie sie beide zusammenwirken.‹ —— **Prof. Dr. Thomas Metzinger, Philosoph, Universität Mainz:** ›Wir werden in der Zukunft die Inhalte unseres bewussten Erlebens wesentlich feinkörniger und genauer kontrollieren können als in der Vergangenheit, indem wir unser Gehirn kontrollieren und technisch beeinflussen. Deswegen ist es wichtig, dass wir uns alle Gedanken darüber machen, was eigentlich ein guter, ein wünschenswerter Bewusstseinszustand ist.‹ —— **Prof. Dr. Günter Dueck, Zukunftsforscher, IBM Distinguished Engineer am Wissenschaftlichen Zentrum, Heidelberg:**

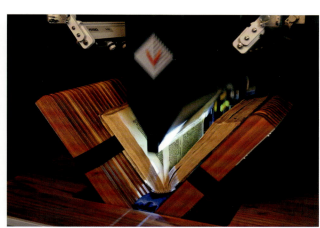

1 Restaurierung von Pergament *Parchment restoration* — 2 Heftbünde *Stitched bindings* — 3 Restaurierung von Buchmalerei *Restoration of manuscript illuminations* — 4 Digitalisierung durch Scanroboter *Digitisation by scan robots* — 5 Herzogin Anna Amalia Bibliothek, Weimar *Duchess Anna Amalia Library, Weimar*

1–4 Institut für Buch- und Handschriftenrestaurierung, Bayerische Staatsbibliothek, München 5 Klassik Stiftung Weimar

›Das Denken schleicht so ein bisschen der technologischen Entwicklung nach, und es wäre schöner, wenn es vorauseilen würde.‹ —— **Prof. Dr. Marion A. Weissenberger-Eibl, Zukunftsforscherin, Fraunhofer Institut für System- und Innovationsforschung (ISI), Karlsruhe:** ›In der dünnen Luft des globalen Wettbewerbs nimmt die Gefährlichkeit extrem unwahrscheinlicher Ereignisse exponentiell zu, und diese Ereignisse entziehen sich ganz einfach der Vorausschau. Dagegen hilft nur, sich dem Unerwarteten mit Neugier anzuvertrauen, denn ich kenne keine bemerkenswerte Technologie, die Resultat einer absichtsvollen Planung ist. Insofern sollten sich Entscheider trainieren, neben der Top-down-Planung viel Raum für maximales Ausprobieren und für das Erkennen sich bietender Chancen zu reservieren.‹ —— **Prof. Dr. Klaus Töpfer, Umweltminister a. D., Berlin:** ›Die Auswirkungen menschlichen Handelns werden nicht erforscht, damit wir sie hinterher so erleben. Sie werden erforscht, damit wir Hinweise darauf bekommen, wie wir die Zukunft gestalten können, sodass sie auch kommenden Generationen ein menschenwürdiges Leben gestatten.‹ —— **Sven Giegold, Wirtschaftswissenschaftler, Verden/Aller:** ›Wir brauchen einen globalen Grünen New Deal, so wie ihn der UN-Generalsekretär Ban Ki-moon fordert: Das bedeutet massive internationale Investitionen in erneuerbare Energien und Energieeffizienz, woraus die Basis für eine neue industrielle Revolution entstehen kann, die uns Arbeitsplätze und gleichzeitig die Überwindung der Klimakrise bringen wird.‹

Internet-Nutzer 1990 *Internet users 1990*

Internet-Nutzer 2002 *Internet users 2002*

The future of mankind

Human consciousness is the product of natural evolution. However this evolution is being perpetuated by man using the artificial resources provided by science and technology. Genes, neurones, memes, laughter and ›pointing‹ renew the perspectives of our self-knowledge. Genetics, cognitive research and information technologies are intervening in the future development of mankind; the acquisition of knowledge and education of the individual will change, as will society. Genetic technology is helping to explain the history of mankind and everyone will soon have access to information about their own individual genome. Reproductive medicine helps childless people to have families. However, the question remains: Where do the ethical boundaries lie and what opportunities does the future actually offer. Scientists, scholars and artists provide answers to the central question: ›What is man?‹ **Prof. Svante Pääbo, paläoanthropologist, Max Planck Institute for Evolutionary Anthropology, Leipzig:** ›When we compare the Neanderthal genome with that of man, we will find the genetic basis for what makes humans unique. What is it in our biology that enables us to develop technology that changes so quickly, to develop language, to develop art and colonise the entire world?‹ **Prof. Hans Lehrach, molecular geneticist, Max Planck Institute for Molecular Genetics, Berlin:** ›In the area of cancer treatment, in particular, we are already in a position to use the knowledge of the genome, the tumour and the patient to administer the existing forms of treatment in a far more targeted way. This approach could also become important in many other areas of life. It would not surprise me if we end up assessing the remedies to be used to prevent hair loss in a particular individual or optimising training programmes at the gym based on the individual genome.‹ **Dr. Monika Bals-Pratsch, expert of reproductive medicine, University of Regensburg:** ›Artificial fertilisation will certainly never replace natural conception, but there are many couples who, for medical reasons, have to avail themselves of the options provided by artificial fertilisation to have their own children, and who then decide to make a donation of semen or eggs or to act as surrogates. Sperm banks and egg banks exist and this gives couples of the same sex, for example, the possibility to have their own children, to establish a family, a so-called rainbow family.‹ **René Röspel, biologist, Member of the German Bundestag, Berlin:** ›More than almost any other country, Germany is dependent on basic research and technical innovation. Of course, innovation, science and technology influence society and ethical and moral issues.‹ **Prof. Niels Birbaumer, neuropsychologist, University of Tübingen:** ›I believe that in the very near future human beings and the human brain and a connected computer will be able to form a kind of symbiosis.‹ **Prof. Thomas Christaller, expert of artificial intelligence, University of Bielefeld and Institute Director at the Schloss Birlinghoven Campus:** ›I find a scientific vision that was defined in recent times very interesting, namely the idea that in the year 2050, a team of humanoid robots could potentially beat the current football world champions – an enormous challenge. If this will be possible – and that remains an open question – these robots will surely engage in all kinds of behaviour, ranging from fair play to the most dangerous sliding tackle.‹ **Prof. Hannelore Ehrenreich, medic, University of Göttingen and Max Planck Institute for Experimental Medicine, Göttingen:** ›Of course, the question arises as to whether erythropoietin could also be used for brain doping in healthy people. Needless to say, this is something I cannot endorse as a scientist, as I assume that the stimulation of brain functions in healthy people follows completely different rules than in sick people. Where we achieve an improvement in patients, we could possibly cause impairment in healthy people.‹ **Prof. Michael Pauen, philosopher, Humboldt University Berlin:** ›I believe that these developments [of neuro science] actually do provide us with instruments that we are capable of making good use of, but which we can also misuse. If we use them properly, these tools will enable us to fight disease better and, for example, overcome inborn performance deficits. If we abuse them, something like brain doping arises, and tendencies akin to those we can observe in the area of cosmetic surgery today, will also emerge in the area of the brain.‹ ——— **Prof. Wolf Singer, brain researcher, Max Planck Institute for Brain Research, Frankfurt:** ›Thus we are subject to the illusion of being more free than we actually are. However, we are still responsible for everything we do because nobody, apart from ourselves and, hence, our brains, is responsible for these decisions.‹ ——— **Prof. Julian Nida-Rümelin, philosopher, Ludwig Maximilian University Munich:** ›What neuro science can't show is that the weighing up of reasons does not play a role in our convictions and our actions – we have as yet no empirical evidence pointing in this direction. If this were the case, the project of science would actually be obsolete.‹ ——— **Prof. Alfred Nordmann, philosopher, Technical University Darmstadt:** ›The distinctive feature of nanotechnology is that the things it talks about are unbelievably, unimaginably small. The uncanny thing is that the technology is disappearing from the horizon of our experience and is becoming something magical and quasi-natural.‹ ——— **Prof. Gerd Karl Binnig, physicist, Honorary Professor at the Ludwig Maximilian University Munich:** ›When we embarked on the development of scanning tunnelling microscopy over 25 years ago, the notion of playing with the atoms seemed relatively normal and almost real to me. If you believe firmly in something, it almost becomes reality. But when I see the relaxed way in which people deal with atoms and this atomic world today, it seems almost unreal to me. So I would say that this relaxed attitude is the real surprise.‹ ——— **Prof. Wolfgang Wahlster, computer scientist, German Research Center for Artificial Intelligence in Saarbrücken, Kaiserslautern, Bremen and Berlin, and University of Saarland:** ›Intelligent software systems are getting better at understanding the multimedia content of the Web. A quasi semantic Web is emerging as a result. Access is becoming multi-modal, the computer is becoming a dialog partner and access is quite intuitive for everyone.‹ ——— **Prof. Gerhard Schulze, sociologist, University of Bamberg:** ›If we look back on the several-hundred-year success story of the modern age, we must admit that this system has worked extremely well. However, it also has its absurd aspects, as beyond all of this striving, this multiplication of possibilities and this desire to attain the best of all worlds, we are always dissatisfied and have lost a particular capacity, without which striving is completely pointless: that is the ability to linger in the wider space of possibility. This means: to be somewhere, to have time, to no longer want to strive, to want to make use of what you have already achieved and take pleasure in it.‹ ——— **Prof. Elsbeth Stern, researcher on learning, ETH Zurich, Switzerland:** ›The fact that we cannot learn without a brain does not mean that brain researchers can actually improve the school system. They are completely unable to do this at present. There is no single result of brain research that would force us to rethink practices at school level. What has emerged from the empirical research on learning carried out by educational science and psychology goes far beyond the trivial statements made by some brain researchers.‹ ——— **Prof. Ursula Staudinger, psychologist, Jacobs University Bremen:** ›Learning is a process that is fundamentally linked with human existence. As long as we live, and do not succumb to any drastic pathological processes, we can learn.‹ ——— **Prof. Aleida Assmann, literary scholar, University of Konstanz:** ›Today, we can hear all of Händel's operas, all of the Beatles' songs, in any place and at any time. We have access to the cultural memory, to its inventories, and can also collect them ourselves. This is an effect of a major process of democratization. However, there is also a catch: these new media are no longer

Probanden aus der Lachforschung (Gelotologie) *Test subjects in laughter research (gelotology)*

Institut für Biologie, AG Humanbiologie und Anthropologie, FU Berlin

1

2 3
4 5

as stable as the old ones and require enormous technical effort and investment for their constant renewal.‹ —— **Prof. Hermann Parzinger, archaeologist, President of the German Archaeological Institute, Berlin:** ›In Europe and many other parts of the world, we are already living in a multicultural society. Many people with migrant backgrounds influence the life in our cities; they bring their history with them. I think it is very, very important for us to acknowledge this history and to better understand this culture.‹ —— **Volker Schlöndorff, film director, Potsdam:** ›Traditional dramaturgy is, of course, also being challenged by brain research, by the question as to whether we actually have free will at all. Because, if I do not have free will, many conflicts have to be resolved in a very different way, stories have to be told in a different way.‹ —— **Prof. Carsten Niemitz, expert of human biology, Free University Berlin:** ›Real laughter is actually an expression of the human being. We inherited it in the course of evolution. Although anthropoid apes can also laugh, in laughing we express empathy, the ability to identify with others in shared amusement. In the course of evolution, we also inherited the ability to dismiss and exclude others through laughter. Thus, in laughter, we have the possibility of establishing social boundaries.‹ —— **Prof. Dieter Hattrup, theologian, University of Paderborn:** ›The old opposition between science and religion is over for me. What matters is how they interact with each other.‹ —— **Prof. Thomas Metzinger, philosopher, University of Mainz:** ›In future we will be able to control the content of our conscious experience with greater precision and in far greater detail than in the past, in that we will be able to control and technically influence our brain. Therefore, it is important that we all reflect on what is a good and desirable state of consciousness.‹ —— **Prof. Günter Dueck, future researcher, IBM Distinguished Engineer at Wissenschaftliches Zentrum, Heidelberg:** ›Thinking always lags somewhat behind technological development and it would be nicer if it were ahead of it.‹ —— **Prof. Marion A. Weissenberger-Eibl, future researcher, Fraunhofer Institute for Systems and Innovation Research (ISI), Karlsruhe:** ›However, the risk of extremely unlikely events is increasing exponentially in the thin air of global competition, and these events simply evade all forecasting. All that helps here is to commit to the unexpected with curiosity; I am not aware of any remarkable technology that is the outcome of intentional planning. Thus, in addition to top-down planning, decision-makers should train themselves to reserve a lot of space for the maximum testing of things, and for the recognition of opportunities that simply arise.‹ —— **Prof. Klaus Töpfer, German Federal Minister for the Environment (retired), Berlin:** ›These effects caused by human action are not being researched so that we can experience them in retrospect, but with a view to obtaining indicators as to how we can shape the future in a way that will enable coming generations to lead a decent life.‹ —— **Sven Giegold, economist, Verden/Aller:** ›We need a global Green New Deal as urged by UN Secretary General Ban Ki-moon. This involves massive investment in renewable energy and energy efficiency throughout the world. The basis for a new industrial revolution will more or less emerge from this, providing both jobs and a solution to the climate crisis.‹

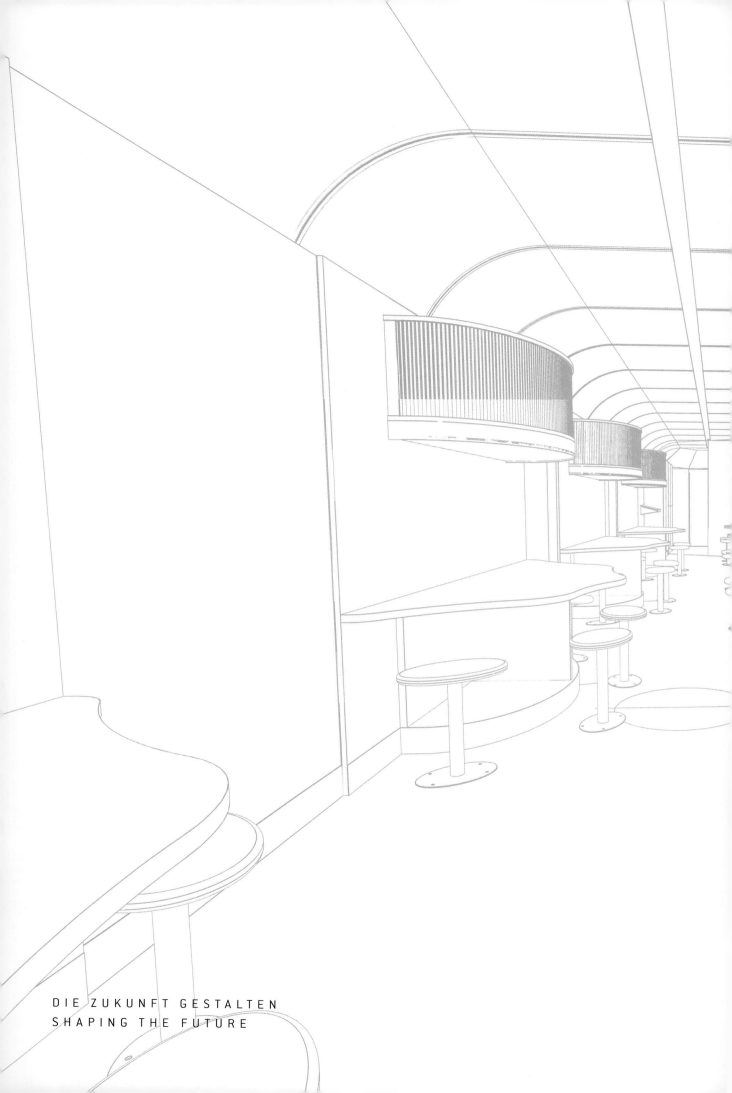

DIE ZUKUNFT GESTALTEN
SHAPING THE FUTURE

Vorige Doppelseite: Neopor, eine Weiterentwicklung von Styropor.
Preceding spread: Neopor, an improved form of styrofoam.
BASF SE

Die Zukunft gestalten

Es genügt nicht zu wissen, man muss es auch anwenden. Es genügt nicht zu wollen, man muss es auch tun. *Johann Wolfgang von Goethe* ▁▁▁ Der Weg von der Idee zur Innovation hängt nicht allein vom wissenschaftlichen und technologischen Umfeld ab. Wir selbst und unser Tun entscheiden darüber, welche Erfindungen und Innovationen unser Leben begleiten. Wissenschaft und Technik liefern für globale wie lokale Herausforderungen wichtige Antworten und Lösungsvorschläge. Welche Richtung die zukünftige Entwicklung in einer globalisierten Welt nimmt, wird geprägt sein vom effektiven Zusammenspiel zwischen Wirtschaft, Gesellschaft, Wissenschaft und Politik bei der Nutzung des neuen Wissens. Neue Ideen werden erst dann in einem Land Wirklichkeit, wenn dieses selbst an deren Entstehung mitgewirkt hat, wenn sie von der Gesellschaft wohlwollend angenommen, von der Politik in die richtigen Bahnen gelenkt und von der Wirtschaft verantwortungsvoll umgesetzt werden. ▁▁▁ **Erstes Mitmachlabor auf deutschen Schienen** Grundlage aller Innovation sind Neugier und Entdeckergeist. Diese sind bei Kindern und Jugendlichen besonders ausgeprägt. Im letzten Wagen des Ausstellungszugs befindet sich deshalb ein Mitmachlabor für die Forscher von morgen. Hier konnten Schulklassen und Familien erleben, wie kreatives Experimentieren zu innovativen Produkten und Problemlösungen führen kann. In den Workshops *Cooler Kunststoff – ein Werkstoff mit Überraschungseffekt* und *Heiße Zellen – Solarenergie für kreative Anwendungen* wurden anhand einfacher, aber eindrucksvoller Experimente die Arbeits- und Denkweisen moderner Forscher vermittelt. Bis zu vier Schulklassen konnten sich pro Tag aktiv mit Wissenschaft und Technik beschäftigen. Nachmittags und an schulfreien Tagen stand das Mitmachlabor vor allem Familien mit Kindern offen.

Shaping the future

It is not enough to know, we must also apply; it is not enough to will, we must also do. *Johann Wolfgang von Goethe* ▁▁▁ The path from inspiration to innovation does not depend on the scientific and technological environment alone. The decision as to which inventions and innovations will become part of our lives is down to us and what we do. Science and technology provide important suggestions and answers to both global and local challenges. The direction to be taken by future developments in a globalized world will be shaped by the effective interplay between business, society, science and politics in the exploitation of this new knowledge. New ideas become reality in a country only when the people there have contributed to their development, when these new concepts are favourably accepted by society, are steered in the right direction by the political system, and are responsibly implemented by the business world. ▁▁▁ **First hands-on laboratory on german tracks** Curiosity and the spirit of invention are the basis of all innovation. These characteristics are particularly prominent among children and young people. For this reason, an interactive hands-on laboratory has been provided in the final carriage of the exhibition train for the researchers of tomorrow. School classes and families could experience here how creative experimentation can lead to innovative products and solutions. The working methods and thought processes of modern researchers have been conveyed in the *Hot Cells* and *Cool Plastics* workshops using simple but impressive experiments. Up to four school classes could actively engage in science and technology. In the afternoons and on school holidays the hands-on laboratory was open to families with children.

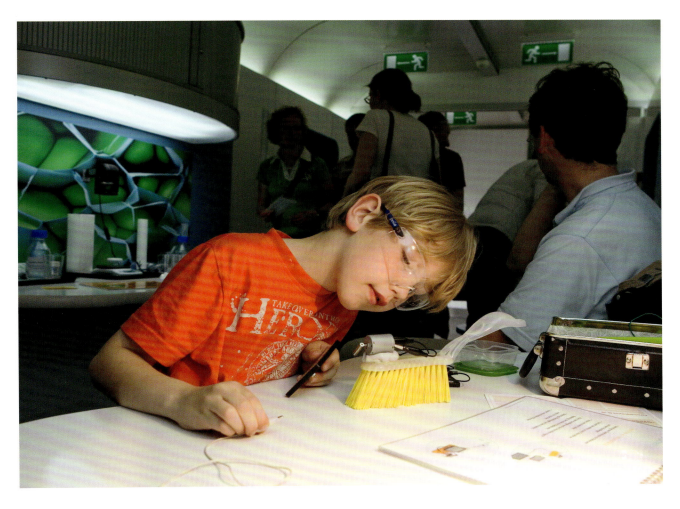

ZAHLEN UND FAKTEN EINER AUSSERGEWÖHLICHEN TOURNEE
FACTS AND FIGURES OF A REMARKABLE TOUR

sie waren am zug!

Ganze 62 Städte Deutschlands, darunter alle Landeshauptstädte, besuchte die *Expedition Zukunft* während ihrer Tournee durch die Bundesrepublik im *Wissenschaftsjahr 2009*. Im 333 Meter langen Sonderzug mit einer Ausstellungsfläche von rund 1.400 m² in zwölf Wagen wurde mit 480 Tonnen nicht nur wissenschaftlich einiges bewegt. Zwei eigens für diesen Zweck gestaltete Elektroloks sowie bei Bedarf auch eine Diesellok zogen den Ausstellungszug kreuz und quer durch deutsche Lande, von Konstanz bis Kiel, von Aachen bis Görlitz. ⎯⎯ Von den 250.000 erwarteten Besuchern konnte der 50.000ste bereits am Pfingstwochenende Ende Mai in Bonn begrüßt werden. Es folgten der 100.000ste Besucher in Nürnberg Mitte Juli und der 200.000ste Anfang Oktober in Potsdam. Einen wahren Ansturm interessierter Bürgerinnen und Bürger auf den Wissenschaftszug verzeichneten besonders die Stationen in den östlichen Bundesländern. An einem einzigen Tag zählte das Zugteam in Berlin 2.857, in Erfurt 2.348 und in Dresden 2.133 Besucher. Bei einer maximalen Kapazität von 300 Personen pro Stunde war die Ausstellung damit vollständig ausgelastet – es kam teilweise zu Wartezeiten von bis zu zwei Stunden und Menschenschlangen von über 100 Meter Länge. Das war auch an den erfolgreichsten Standorten der Fall: In Dresden kamen insgesamt 5.742 und in Erfurt 5.606 Besucher, dicht gefolgt von Aachen mit 5.317 und Wuppertal mit 5.304 Gästen. ⎯⎯ Auch viele hochrangige Vertreter aus Politik, Wirtschaft und Wissenschaft ließen es sich nicht nehmen, das größte vom Bundesministerium für Bildung und Forschung geförderte Projekt im *Wissenschaftsjahr 2009* zu bestaunen. Insgesamt kamen mehr als 1.500 geladene Gäste deutschlandweit zu den Begrüßungsveranstaltungen, mehr als 300 mit herzlichen Grußworten für die *Expedition Zukunft* – mancherorts mit musikalischer oder künstlerischer Untermalung. So machten zwei DJs den Halt in Jena zu einem Sommerfest, spielte in Ulm eine Jazzkapelle auf, und in Rostock fand sogar eine Tanzperformance zum Empfang der *Expedition Zukunft* am Bahnhof statt. ⎯⎯ Um den täglichen Betrieb zu gewährleisten, wurden vor Ort insgesamt mehr als 350 Hilfskräfte zur Unterstützung des Zugteams eingesetzt. Denn das feste Personal war täglich damit beschäftigt, insgesamt mehr als 1.800 Führungen und 382 Workshops zu veranstalten. Auch die Organisation des Tagesablaufs, besonders im Labor, war gespickt mit Superlativen: Fast 40.000 Besucher schauten im Labor vorbei, um einen oder beide Workshops zu absolvieren. Dabei brachten sie ›Krabbler‹ aus verschiedenen Bürsten mithilfe einer künstlichen Sonne und ›Heißen Zellen‹ zum Laufen: 50 cm in 3,6 Sekunden lief die Froschbürste, Weltrekord! Auch beim ›Coolen Kunststoff‹ fielen Rekorde: Eine Superabsorberkugel konnte in Heidelberg das 206-fache ihres Gewichts an Wasser aufnehmen. Damit passten dann nur noch 33 Kugeln in einen Becher, statt 7.103 im trockenen Zustand. ⎯⎯ Nach der großen Deutschlandreise zeigten die Kilometerstände der drei Teamtransporter, mit denen das Team dem Zug von Stadt zu Stadt folgte, jeweils rund 20.000 km. Diese Kilometerzahl wird die *Expedition Zukunft* voraussichtlich auch 2010 auf ihrem Weg durch die Volksrepublik China erreichen.

During its tour of Germany in the *Year of Science 2009*, the *Science Express* visited a total of 62 German cities, including all of the federal state capitals. Science was not the only thing on the move in the 333-metre-long special train, which had an exhibition area of around 1,400 m² distributed across twelve carriages and totalling 480 tonnes in weight. Two specially-designed electric locomotives and if required one diesel locomotive pulled the exhibition train all over Germany, from Constance in the South to Kiel in the North and from Aachen in the West to Görlitz in the East. ⎯⎯ Of the 250,000 expected visitors, the 50,000th was greeted as early as Whitsun weekend at the end of May in Bonn. The 100,000th visitor followed in Nuremberg in mid-July and the 200,000th in early October in Potsdam. The stations in the eastern federal states, in particular, experienced a veritable rush of interested citizens. The highest numbers of visitors recorded on a single day were 2,857 in Berlin, 2,348 in Erfurt and 2,133 in Dresden. At a maximum capacity of 300 people per hour, this meant that the exhibition was constantly full: waiting times of up to two hours arose in some places and queues formed of more than 100 metres. This was also the case in the most successful locations: the record number of visitors was achieved by Dresden and Erfurt with 5,742 and 5,606 visitors respectively, followed closely by Aachen with 5,317 and Wuppertal with 5,304. ⎯⎯ Many high-ranking representatives from politics, business and science could not resist the opportunity to admire the biggest project funded by the Federal Ministry of Education and Research in the *Year of Science 2009*. A total of over 1,500 invited guests from all over Germany came to the individual launch events, more than 300 of them greeting the *Science Express* in the warmest of terms in their welcome addresses, in some locations with background music or artistic accompaniment. For example, two DJs created a summer festival atmosphere for the train's stopover in Jena; the arrival of the train was accompanied by a jazz band in Ulm, and in Rostock it was welcomed by a dance performance. ⎯⎯ A total of over 350 temporary staff were employed on-site to support the train team and ensure the smooth daily operation of the train. The permanent staff were kept busy every day with over 1,800 tours and 382 workshops. The organisation of the daily schedule, particularly in the laboratory, was full of superlatives: Almost 40,000 visitors stopped by at the laboratory to take part in one or both workshops. This involved getting ›crawlers‹, made of various brushes, to walk with the help of an artificial sun and ›hot cells‹: 50 cm in 3.6 seconds moved the frog-brush, a world record! Also in the ›cool plastics‹ section records were set: a super absorber ball created in Heidelberg was able to absorb 206 times its own weight in water. As a result only 33 balls could fit in a cup rather than 7,103 in their dry state. ⎯⎯ After the long journey throughout Germany, the three team transporters, with which the team followed the train from city to city, had clocked around 20,000 km each. A distance that will presumably be covered by the *Science Express* also on its travels through the People's Republic of China in 2010.

So stelle ich mir die Zukunft vor In ein elektronisches Gästebuch oder handschriftlich auf der Broschüre zur *Expedition Zukunft* haben die Besucher Vorstellungen von einem zukünftigen Leben eingetragen. Entstanden ist dabei ein breites Spektrum an Ideen, aber auch an Zweifeln über unsere Zukunft.

This is my idea of the future In the electronic guest book or handwritten on the *Science Express* brochure, the visitors have expressed their vision of life in the future. A broad range of ideas developed, but also scepticism emerged about where our world is heading.

Etwas Angst, als älterer Mensch mit den Neuerungen überhaupt noch zurecht zu kommen. **BARBARA, 68 JAHRE** *As an older person, I am rather afraid that I may not be able to cope with the innovations.* **BARBARA, 68 YEARS**

Ich erwarte von der Zukunft, dass alles viel einfacher wird. Zum Beispiel könnte man den Schülern statt 6–8 Stunden Schule, einfach einen Helm aufsetzen, der alles Wissen in die Gehirne einspeichert. **ANONYM, 13 JAHRE** *What I expect from the future is that everything will be much simpler. For example, instead of having to attend school for six to eight hours, pupils could simply have a helmet placed on their heads that will feed all the knowledge into their brains.* **ANONYMOUS, 13 YEARS**

Diese großartigen und einprägsamen Darstellungen und Instruktionen sollten viele oder besser alle Erdenbürger dazu bewegen, sparsamer mit den Schätzen unserer Erde umzugehen. Junge Menschen sollten dazu viel intensiver belehrt und ausgebildet werden. Lernwillig sind sie und sollten es bleiben!! **JOACHIM, 95 JAHRE** *These marvellous and impressive presentations and instructions should move many or, better still, all citizens of the Earth to be more careful in dealing with the treasures our planet has to offer. Young people should be given far more intensive education and training for this. They want to learn, and should stay that way!!* **JOACHIM, 95 YEARS**

Verschärfung der Gegensätze von arm + reich. Keine ausreichende Reaktion auf den Klimawandel und den damit einhergehenden Folgen. **EIKE, 49 JAHRE** *Widening of the gap between rich and poor. Insufficient response to climate change and the resulting consequences.* **EIKE, 49 YEARS**

Bessere Mittel gegen Pickel. Schuldenerlass für die ausgebeutete 3. Welt. Freies offenes Wissen für alle Erdenbürger. **SUSANNE, 24 JAHRE** *Better treatments for spots. Debt relief for the exploited Third World. Freely-available knowledge for all of the Earth's inhabitants.* **SUSANNE, 24 YEARS**

Roboter, die den ganzen Haushalt machen. Ein Roboter, der für mich in der Schule nur Einsen schreibt. **FERDINAND, 10 JAHRE** *Robots that do all the housework. And a robot that would get top grades for me at school.* **FERDINAND, 10 YEARS**

Deutschland als Vorbild für Demokratie und Frieden. Eine Besinnung auf die Werte und [...] die Erde schätzen und achten. Jeder kann im Kleinen viel dafür tun: Energie sparen und Nächstenliebe üben, einen Glauben auch an die Zukunft entwickeln. **VALESKA, 41 JAHRE** *Germany as a model for democracy and peace. An awareness of values and [...] treasuring and respecting the Earth. Everyone can do a lot for this in small ways: saving energy, practising brotherly love, developing faith in the future as well.* **VALESKA, 41 YEARS**

Ich erwarte in einer noch weiter digitalisierten Welt zu leben, in der unsere Entscheidungen auf neuen Technologien und nicht unseren eigenen Intuitionen beruhen. Ich erwarte und erhoffe mir eine gemeinsame Lösung für soziale und politische Grenzen für eine gerechtere Welt, Gleichberechtigung und gleiche Chancen für alle. Ich erwarte wissenschaftsbasierte Veränderungen beim Verkehr und Automobiltechnologien, und dass der Untergrund stärker genutzt wird als heute. **ANONYM** *I expect to live in a more digitalised world where human decisions will depend on new technologies, not individual intuitions. I expect and hope for a common solution to social and political boundaries, for a fairer world, equal rights and opportunities. I expect scientific changes in transport and car technologies, and that the underground will be used more then now.* **ANONYMOUS**

Ich will Superkräfte haben. Ich will, dass wir uns im Weltraum ansiedeln. Ich will, dass wir statt Autos Raketenrucksäcke benutzen. Ich will, dass ich eine Armee Klonkrieger besitze. **ELIAS, 9 JAHRE** *I want to have super powers. I want us to inhabit space. I want us to use rocket rucksacks instead of cars. I want to own an army of clone warriors.* **ELIAS, 9 YEARS**

Ich habe Angst vor der Zukunft, insbesondere wegen der atomaren Energie und des Klimawandels. **VON ANONYM** *I am afraid of the future, in particular because of nuclear energy and climate change.* **ANONYMOUS**

Ich hoffe, dass man in Zukunft Krankheiten wie Krebs heilen kann. Außerdem hoffe ich, dass das Sicherheitssystem besser ist als heute, damit es nicht mehr so viele Terroranschläge gibt und ich hoffe, dass sich die geschützten Tierarten erholen können und die Bestände dieser Tierarten wieder steigen. **JONAS, 13 JAHRE** *I hope that we will be able to cure diseases such as cancer in the future. I also hope that the security system will be better than today so that there are not as many terrorist attacks, and that the protected animal species can recover and the stocks of these species increase again.* **JONAS, 13 YEARS**

Ich erwarte mir von der Zukunft, dass man durch Solarzellen mehr Energie gewinnen kann. Außerdem sollte man die Klimaerwärmung stoppen, dass Eisbären besser leben können und so nicht aussterben. Das könnte durch eine Verringerung des CO_2-Ausstoßes verändert werden. **ANONYM** *I expect from the future that we will be able to obtain more energy from solar cells. Also, climate warming should be stopped so that polar bears can live better and do not become extinct. This could be changed by reducing CO_2 emissions.* **ANONYM**

Saubere Luft durch umfunktionierte Schornsteine von Fabriken, die Schmutzpartikel ansaugen. **SILKE, 42 JAHRE** *Clean air through converted factory chimneys which suck up the dirt particles.* **SILKE, 42 YEARS**

Sparen Sie sich das künstliche Gehirn zu erzeugen: Heilen Sie Krankheiten! Sparen Sie sich nach einer zweiten Erde zu suchen: Kümmern Sie sich um unsere! Vernetzte Welt, wo man von zu Hause in jeder Firma auf der Welt arbeitet. Forscher, je nach Bedarf zerstreut auf alle Kontinenten und zu einem Institut gehörend. **ASMA, 9 JAHRE** *Don't bother developing the artificial brain: heal diseases! Don't bother looking for a second Earth: focus on the one we have! A networked world, where you can work for any company in the world from home. Researchers spread across all continents as needed and belonging to one institute.* **ASMA, 9 YEARS**

Ich wünschte mir, es gäbe eine Brille, wo ganz viele Bücher drauf gespeichert wären. **VON ANONYM** *I wish there would be glasses in which many books could be stored.* **ANONYMOUS**

ist 300 Meter lang

MP Wulff begeistert: "Forschung wird greifbar."

"Wie im Science-Fiction-Film"

"Zug der Wissenschaft" im Hauptbahnhof. Computer erkennen Emotionen.

VON KAREN VON SCHMIEDEN

Richtung Zukunft

Zukunft auf Rädern

Wissenschaftszug macht Station in Dortmund: Ab Sonntag auf Gleis 26

... Zug gelöscht

Der Wissenschaftszug macht Station in Nürnberg

Warten auf die Zukunft

Lange anstehen am Zug der Zukunft

Riesiger Ansturm – die Wissenschafts-Schau bleibt noch bis morgen

...derzug in die Zukunft

...e rollende Forschungsshow am Frankenstadion

...ende Reise im stehenden Zug

Zugkräftig: 300 Meter geballte Wissenschaft

Mobile Ausstellung "Expedition Zukunft" macht von Dienstag bis Donnerstag im Hauptbahnhof Station

Nimm doch den Wissenschaftszug

Seit gestern macht die Ausstellung »Expedition Zukunft« Halt in Freiburg / Lange Warteschlangen

KONZEPT UND REALISIERUNG *concept and realization*

Der Ausstellungszug *Expedition Zukunft* ist ein Projekt der Max-Planck-Gesellschaft. Dieses knüpft an die Erfahrungen der international erfolgreichen Wanderausstellung *Science Tunnel* sowie an den Ausstellungszug *Science Express* in Indien an. The exhibition train *Science Express* is a project of the Max Planck Society. It is based on experiences of the internationally successful travelling exhibition *Science Tunnel* and the exhibition train *Science Express* in India.

PROJEKTLEITUNG UND GESAMTKONZEPTION *leading project manager and overall concept* Dr. Andreas Trepte
PROJEKTMANAGEMENT *project management* Dr. Peter M. Steiner
NATUR-, MATERIALWISSENSCHAFTEN, TECHNIK *natural and materials science, technology* Dr. Hannelore Hämmerle, Dr. Christoph Ettl (Wissenschaftliche Beratung und Konzeption), Helmut Hornung
BIOLOGIE, MEDIZIN, LANDWIRTSCHAFT *life science, medicine, agriculture* Nadja Pernat, Dr. Christiane Walch-Solimena (Wissenschaftliche Beratung und Konzeption)
WISSENSCHAFTSTHEORIE, KULTURWISSENSCHAFTEN *philosophy of science, cultural studies* Dr. Peter M. Steiner, Dr. Andreas Trepte
AUSSTELLUNGSINFORMATIONSSYSTEM, PROJEKTASSISTENZ, ADMINISTRATIVE KOORDINATION *exhibition information system, project assistance, administrative coordination* Jan Bejenke
WEBSITE, BILDBETREUUNG *website and multimedia support* Dalija Budimlic
IT-UNTERSTÜTZUNG *IT support* Julius Benkert

ZUGTEAM DER MAX-PLANCK-GESELLSCHAFT *train crew of the Max Planck Society*

AUSSTELLUNGSMANAGEMENT, -PERSONAL UND -BETRIEB *exhibition management, staff and operations* Max-Planck-Gesellschaft, Projektteam Expedition Zukunft
AUSSTELLUNGSLEITUNG VOR ORT *exhibition management on site* Dr. Hannelore Hämmerle, Jörg Theurer
TECHNISCHER ZUGLEITER *technical train management* Josef Binder
AUSSTELLUNGSTECHNIKER *exhibition technician* Peter Krause
ZUGPERSONAL *guides and presenters* Jutta Böhme, Albia Consul, Yvonne Dieckhoff (Mitmachlabor), Claudia Dobrinski, Dr. Jan Fischer, Carolin Hanke (Mitmachlabor), Andreas Hecker, Meike Jotzo, Klaus von Kittlitz, Sophie Kolb, Lisa Krassuski, Tobias Lauterbach (Mitmachlabor), Nora Lessing, Heidrun Mader, Tine de Maeyer, Dr. Karen Pfister (Mitmachlabor), Ulrike Richter, Dorte Riemenschneider, Michelle Röttger, Dr. Doris Schmidt, Dr. Cornelia Schmutz (Mitmachlabor), Dr. Clemens Schneeweiß, Klaus Schuler, Elisabeth Steindl
AUSSTELLUNGSFOTOGRAFIE IM ZUG *exhibition photography on the train* Ulrike Richter, Meike Jotzo, Dr. Clemens Schneeweiß

MEDIENARBEIT *media and public relations* Iserundschmidt, Berlin und Bonn

EVENTORGANISATION *event organisation* IMAGO GmbH, Dortmund

MITMACHLABOR *hands-on laboratory* two4science GmbH

GESTALTUNG UND REALISATION *design and realisation*

ArchiMeDes GbR

PROJEKTLEITUNG *project management* Jörg Schmidtsiefen, Werner Rien
AUSSTELLUNGSDESIGN *exhibition design* Michael Feser, Christine Krüger, Gunnar Mikael Gräslund
REDAKTION *editorial staff* Dr. Antonia Humm, Ursula Schmidt, Anna Schäfers
SEKRETARIAT *office* Janine Albrecht, Zdenka Karacic, Melanie Schonert

GRAFIKDESIGN *graphic design*

ART DIREKTION *art direction* Günther Albien
CORPORATE DESIGN *corporate design* Stefan Hofer
GRAFIK *graphics* Tania Hartmann
TYPOGRAFIE, KATALOGGESTALTUNG *typography, catalogue design* Tilmann Benninghaus
PRODUKTIONSLEITUNG *production management* Adam Naparty
GRAFIKASSISTENZ *graphics assistance* Nina Farsen, Ole Bahrmann, Daniel Scheidgen
KORREKTORAT *proofreading* Diedrich Ausprunk

MEDIENDESIGN *media design*

SCREEN- UND MOTIONDESIGN *screen and motion design* Harald Neumann, René Bachmann
MEDIENPRODUKTION *media production* Irene Hardjanegara, Daniel Huber
SOUNDDESIGN *sound design* Frank Martinique

PROJEKTSTEUERUNG UND EXPONAT-ENTWICKLUNG *project management and development*

Archimedes Solutions GmbH

PROJEKTLEITUNG *project management* Stephan Spenling, Gunnar Behrens
EXPONATGESTALTUNG *industrial design* Pascal Wiedenmann, Ulrich Merz, Tim Feltz, Martin Bramer, Johannes Köpp, Marie Seebach, Oliver Klein, Frank Spenling, Yvonne Weber, Jorge Amor Prado, Sabrina Henssen
SOFTWAREENTWICKLUNG *software development* Ulrich von Zadow, Marco Fagiolini, Jens Wunderling, Andreas Dietrich, Holger Frey
MEDIENTECHNIK *technical supervision* Michael Kreuzwieser, Andreas Lüken
ELEKTRONIK *electronics* Gerold Saathoff, Martin Lehmann
KONSTRUKTION *construction* Ole Oppermann, Robert Peter Saladin

TBT – Technischer Bühnen- und Theaterservice

PROJEKTLEITUNG *project management* Michael Köhler
TECHNISCHE LEITUNG *technical supervision* Volker Brenner
BAULEITER *construction management* Lars Weber
FINANZCONTROLLING *financial controlling* Natascha Munera
TECHNISCHE ZEICHNER *draughtsmen* Elke Steinbach, Katrin Strauß, Andreas Roth, Martin Brick, Till Sucker, Mathias Hofmann